SB
950
.L42   The Least is best
         pesticide strategy

DATE			

© THE BAKER & TAYLOR CO

THE LEAST IS BEST PESTICIDE STRATEGY

THE LEAST IS BEST PESTICIDE STRATEGY

A Guide to Putting Integrated Pest Management Into Action

Edited by
Jerome Goldstein

Assistant Editor:
Rill Ann Goldstein

The JG Press, Emmaus, PA

Printed in the United States of America on recycled paper.

Illustrations by Daryl Hunter
Book design by T.A. Lepley

Library of Congress Cataloging in Publication Data
Main entry under title:

The Least is best pesticide strategy.

   Bibliography: p.
   Includes index.
      1. Pest control, Integrated—United States.
      2. Pesticides—United States.
I. Goldstein,
    Jerome, 1931–
II. Goldstein,
    Rill Ann, 1954–
SB950.L42   632′94   78–12806
ISBN 0–932424–00–7

Contents

IPM RESEARCH—PAST, PRESENT AND FUTURE

Acknowledgments

This publication cites the work of many researchers and farmers who have spent years in laboratories, trial grounds and farm fields advancing our knowledge of Integrated Pest Management. This book has been shaped by their efforts and their visions.

Members of entomology, agronomy, horticulture and forestry departments at state universities, and individuals with United States Department of Agriculture research stations and extension offices were most helpful in providing data on current research projects and course offerings. The Association of Applied Insect Ecologists gave us insight into the services of private consultants.

Special thanks go to such persons as Helga and Bill Olkowski, Joel Grossman, Ron Prokopy, Richard Ridgway, Waldemar Klassen, Don Cunnion, Rosmarie von Rumker, William Lockeretz, Sara Wernick, W.R.Z. Willey, Darwin C. Hall and Carlo M. Ignoffo for their contributions, and to Mildred Lalik and Ina Goldstein for their invaluable aid in preparing the manuscript.

Jerome Goldstein
Rill Ann Goldstein
May, 1978

Preface

"Up to now, pesticides often have not been used
for pests; they have been for insurance. The farmers and entomologists don't always know the exact
nature of pest problems, so they say, 'spray to be
safe.' With on-line management, however, we
know exactly what is going on and can predict
what will happen next. This allows the farmer to
spray only when necessary, and results in a major
savings in on-farm spray, labor and energy costs, as
well as reducing the hazard of environmental pollution. . . . Just good scientific pest management on
a typical farm could probably cut pesticide use by
half. . . . Pesticides should be a last resort, after all
else has failed."

Dean L. Haynes
Ecosystems Design and Management Program
Michigan State University

Reducing the amount of pesticides used in America and in the
world is one of the great challenges of our time. Fortunately for all
of us, dedicated and talented persons—in laboratories, research plots
and farm fields—are proving how the least (sometimes none at all)
amounts of pesticides can achieve satisfactory control levels. These
men and women are part of a dynamic field known as Integrated Pest
Management. This book describes many of the prime movers in the
field—the state of their art, the state of their businesses and the
prospects for the future.

We have also tried to explain how the problems confronting
Integrated Pest Management (IPM) can be overcome. Currently,
too many missing links exist—between farmers using IPM approaches and consumers desiring IPM-grown foods; between independent consultants and farmers fed up with excessive pesticide
costs; between students seeking a career in the "IPM industry" and
potential employers; between those envisioning a world whose resources are protected and those with the knowledge to make that
dream come true.

Hopefully, this book will help to forge new links, so that Integrated Pest Management becomes a genuinely effective force in
pest control and in cutting pesticide use by half . . . and by half
again . . .

Chapter 1

De-Spraying America

A change of major importance is taking place on America's farms. The change concerns pest control methods for our nation's food and fiber production.

For the past 30 years, pesticides have been considered to be the number one control. Name a pest, and there's a product from a chemical company that can be sprayed once, twice or a dozen times, or in combination with another product. Such regular spray schedules, we are repeatedly informed, serve as an "insurance policy" for an abundant and safe food supply for both the farmer and the consumer.

But the premium costs for that kind of insurance policy are becoming exorbitant. Farmers are caught in the squeeze of higher pesticide costs and lower pesticide performance. Both growers and consumers worry about health hazards from chemical runoff into water supplies, and from spray residues on foods. They are wary of environmental disasters from accidental spills and human errors. Applicators run the additional risk of direct poisoning. Problems mount from pesticide resistance, secondary pest outbreaks and destruction of natural controls and pollinators.

Slowly but surely, a far different solution is being adopted—*Integrated Pest Management.* Definitions of Integrated Pest Management (IPM) may vary slightly, but its key components have been consistently described by experts in the United States Department of Agriculture (USDA), university entomology departments, state experiment stations, the Environmental Protection Agency (EPA) and by people who are in the business of offering pest management services.

The key components are biological, cultural, physical and genetic controls—with heavy emphasis on monitoring pest populations. Instead of using chemical treatment as a first line of crop protection insurance, chemicals drop way down on

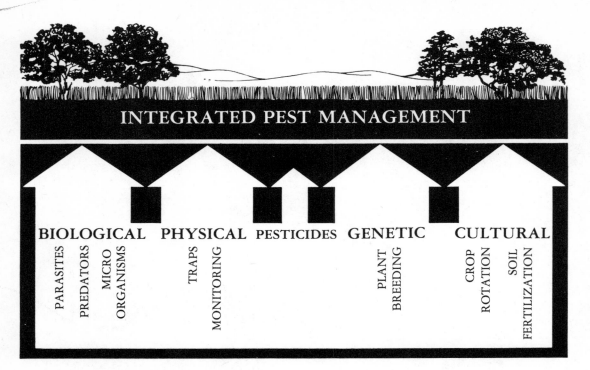

INTEGRATED PEST MANAGEMENT

BIOLOGICAL **PHYSICAL** PESTICIDES **GENETIC** **CULTURAL**

PARASITES
PREDATORS
MICRO ORGANISMS
TRAPS
MONITORING
PLANT BREEDING
CROP ROTATION
SOIL FERTILIZATION

the list, applied only when necessary to prevent economically important damage to the crop.

To control pests with the least amount of chemicals means that people (not products) are needed during the crop season to provide farmers with accurate information. IPM relies heavily on "scouts" going into the farmers' fields to examine crops and monitor the number of plant pests. Monitoring acres of plants is essential in managing pests with an integrated approach. Cotton plants, for example, tolerate high pest populations during certain growth stages. Thus, measuring plant growth becomes as important as sampling insect populations with a sweep net. Each week, hours are spent in fields counting plant parts, taking measurements and jotting down notes. Then all the information is recorded in graphic form to provide the grower with a permanent record of crop growth. The effects of a particular irrigation or cultural practice can be noted from the records, and year-to-year comparisons

can be made.

Scientists evaluate pest counts, along with data on local weather patterns, yields, harvest dates and regional trends. Then, based upon field conditions, farmers receive advice from trained consultants on ways to control crop pests which involve integrated management practices resulting in profitable yields and quality crops. IPM consultants frequently offer advice on fertilization, irrigation, pruning and land use —factors related to pest management which influence crop yield and quality. The per-acre fee is based upon services offered and benefits received.

IPM is an interdisciplinary science which is being advanced by specialists in entomology, plant pathology, nematology, weed science, horticulture, economics, systems science and agriculture. The on-line pest management program at Michigan State University illustrates how it is possible for farmers, consultants, researchers and professors to be informed about the latest developments in IPM. Entomology

Professor George Bird describes the program:

"On-line pest management uses modern communication procedures and data processing technology as a fundamental framework for the program. On-line pest management can be divided into five basic components: a computerized information delivery system, biological monitoring strategies, environmental monitoring networks, appropriate pest-crop ecosystem models, and pest control strategy recommendations.

"Implementation of an on-line pest management program provides many new options. It will also create a number of unexpected problems. One of the major problems that we have faced is getting extension specialists to use the large volume of regional biological and environmental monitoring data on a real-time basis. Specialists have never before had the luxury of such information, and it is frequently difficult to have them change trajectory from their present techniques to a far more precise yet dynamic system."

Just as it is with humans, sex drive is a potent force in establishing insect behavior. Understandably, scientists have begun to use the knowledge of sexual mores to devise environmentally-sounder protective techniques.

Sterilization of males, as in the case of screwworm flies, has been a most important weapon, and artificially-sterilized males are mass-reared at federal facilities in Mission, Texas. In Florida, a Department of Agriculture entomologist has found that the sex scent of each female insect produces an infra-red wave—somewhat like a radio wave, only shorter. He developed a trap which duplicates this electronic radiation and broadcasts it, thereby attracting the males. With a powerful amplifier, entire corn fields can be protected.

Pheromones are increasingly used in IPM programs to attract, confuse, trap and/or monitor insect pests. Pheromone is an American word coined from the Greek *pherein* (to carry) and *hormon* (to stimulate, to excite). The term describes an odorous secretion of an animal or insect that is emitted for the purpose of communicating with other creatures of the same species.

Pheromone traps and spheroid-shaped, sticky-covered traps hanging from tree branches are used in apple and walnut orchards for controlling the codling moth. The scent of the female codling moth attracts male moths to the trap; an IPM practioner keeps a daily record of this number of moths, thereby eliminating at least several sprays used in conventional apple orchard programs.

Professor Ronald J. Prokopy, an entomologist at the University of Massachusetts, is dramatically showing New England commercial orchardists how IPM research findings on pheromones can be utilized on their farms.

Egg-laying by apple maggot females is accomplished when the female arrives on a susceptible fruit, raises up on its legs, bores with its ovipositor through the skin of the fruit into the flesh, and deposits a single egg. The ovipositor is a needle-like protrusion from the posterior of the abdomen through which the egg is passed into the fruit. Following egg deposition, the female withdraws its ovipositor from the fruit, and then proceeds to circle around the fruit for about 30 seconds, dragging its fully-extended ovipositor on the fruit surface behind itself. After this, the female cleans its ovipositor for a few seconds and then flies off the fruit.

Prokopy and some colleagues have

"It is imperative that pest managers be aware of the ecological and biological effects of their actions at all times, not only on the target pest organisms, but also on natural enemies and nonpest organisms."—Mary Louise Flint and Robert van den Bosch.

studied this elaborate behavior, which takes place on cherries and other fruits as well as apples, and identified a fruit-marking pheromone that deters egg-laying. Colleagues in Switzerland mixed this substance in water and applied the solution to cherry trees. The results were extremely encouraging: only six percent of the pheromone-sprayed cherries had any cherry maggot eggs or larvae, compared with 100 percent maggot infestation of adjacent unsprayed cherries.

Along with the identification of the pheromone, Prokopy and an associate also discovered that the fruit-marking pheromone of the apple maggot acts as a chemical signal to *Opius lectus,* a parasite of the maggot eggs. The pheromone arrests the parasite females, and elicits a strong degree of searching behavior for maggot eggs. "Parasites encountering fruits sprayed with fruit-marking pheromone," Prokopy notes, "are therefore likely to remain in the area of the pheromone-sprayed tree for a longer time and effectively search out any maggot eggs that might be in the fruit. If some day the pheromone can be obtained reasonably, then it could be sprayed onto our apple trees to prevent maggot fly egg-laying. The deterred females, which we know move about frequently, might then be captured out by baited yellow rectangles and/or baited red spheres hung in specific trap trees. An integrated approach—combining deterrents, attractants and parasites—could hopefully be achieved."

On such hopes, research and vision will a "least-is-best" pesticide strategy be achieved.

"Integrated pest management systems are dynamic as are the ecosystems in which they are invoked," explain Mary Louise Flint and Robert van den Bosch of the University of California. "In IPM, control action programs evolve as and if pest problems develop; there is none of the preprogrammed rigidity present in conventional pesticide systems where pest managers spray automatically according to a predetermined schedule." They add:

"The integrated pest manager or advisor is the key to this dynamism. He/she must constantly 'read' the situation, evaluate pest populations according to previous experience and knowledge and initiate actions and choose appropriate materials and agents as conditions dictate. It is imperative that pest managers be aware of the ecological and biological effects of their actions at all times, not only on the target pest organisms but also on natural enemies and nonpest organisms."

IPM has its roots in ancient farming and its expression in futuristic computerized technologies. Depending upon one's outlook, it can either be termed old or new. As with anything worthwhile, it is a blend of both—and is as good as the people who implement it.

The list of crops on which IPM has worked is seemingly endless. Carl B. Huffaker of the University of California and B.A. Croft of Michigan State University cite its effectiveness on cotton, apples, alfalfa, corn, pine forests, cereals and citrus. IPM works against many kinds of aphids, bark beetles, mosquitoes, flies, the European corn borer, wheat stem sawfly, codling moth, grape leafhopper, olive knot, sugar beet cyst nematode, potato tuberworm and tomato fruitworm.

The economics of IPM get glowing marks. In California alone, estimated Dr. Paul R. DeBach, a leading IPM figure, producers and consumers of agricultural products have been saved almost $300 million since 1923. Evaluations of IPM programs consistently verify its financial payoff.

Robert Brooks of the University of Florida reports that Florida citrus growers save $10 million per year in reduced insecticide costs from biological controls. As monies now being expended for chemicals to control pests are spent in the information-gathering and interpretation stages of IPM, wide-spread job opportunities and careers are being created.

The environmental benefits of IPM are heralded. Says California entomologist Ray F. Smith: "Based on ecological principles, it integrates multi-disciplinary methodologies in developing agroecosystem management strategies that are practical, effective, economical and protective of both public health and the environment."

With such an outstanding report card, it is surprising that more pesticides than ever are being manufactured and applied this year. Forecasts by market analysts indicate continued growth for the pesticide industry.

The pesticide industry in the United States is a huge business. More than two billion dollars of pest control chemicals are bought each year, with 90 percent used on farms. The balance is divided between industrial use and exterminating companies.

According to a leading New York-based market research organization, Frost & Sullivan, the pesticides market will climb to $3.3 billion by 1984. "Such an optimistic forecast takes into account recently enacted environmental regulations that have stung pesticide producers and users," reports a Frost & Sullivan study entitled *The U.S. Pesticide Industry*. Noting that regulations may force small formulators and manufacturers out of business, the analysts theorize that only the big companies—Monsanto, Stauffer, Eli Lilly and Dow—will be able to afford the expense. "By the next decade, all of the principal pesticide manufacturers are likely to be the pharmaceutical companies."

Herbicides constitute more than 60 percent of total farm use of pesticides, with insecticides and fungicides making up the balance. Looking at the long-term pesticides market, Frost & Sullivan see sales figures this way:

(Millions $U.S.—Users Level)

	1966	1974	1984
Herbicides	243	1,058	1,587
Insecticides	195	491	766
Fungicides	33	116	198

The rapid increase in pesticide use on farms, while widely credited with corresponding to increased crop yields, is also an important factor in decreased farm employment. USDA statistics show that farm employment has fallen from more than eight million workers in 1955 to less than two million today. The drop, in the eyes of USDA policy-makers, was the price that had to be paid for efficiency. A new look is now being given to the term "efficiency," definitely as it relates to pest control methods. Secretary of Agriculture, Bob Bergland, suggested in early 1978 that more people be employed in pest control to conduct surveys and to keep a close watch on pests so they can be controlled more effectively, instead of only using chemicals. He noted that, "many insects have built up resistance to chemicals; the cost of pesticides are up and supplies of petroleum-based products used to make many chemicals are limited."

No One Is Satisfied

Although the pesticide industry is reaping enormous profits, this form of pest management is not profitable nor satisfactory for many people in the United States and elsewhere. Actually, no one is satisfied with the prevalent use of the "most-is-best" pesticide strategy.

For farmers—pesticide costs are increasing, along with the amounts sprayed; inefficiencies are rampant, as pests build up a resistance to chemicals in a vicious cycle; cases of human poisoning, health problems and pollution runoff are multiplying. Cotton growers in the South are in a terrible

bind. In Louisiana, Arkansas, Mississippi and more recently in California as well, farmers are having all kinds of problems controlling the tobacco budworm in cotton fields. Observes James W. Smith, a research entomologist at the Bioenvironmental Insect Control Laboratory in Stoneville, Mississippi:

"Many producers in this area failed to control the tobacco budworm with repeated applications of recommended insecticides with cost approaching $100 per acre. Producers simply cannot tolerate this expenditure for insect control and continue to grow cotton at the present prices. The problem is very complex and has been aggravated by modern agricultural practices. Consequently, other approaches, in conjunction with insecticidal control, are imperative. Good, well-managed Integrated Pest Management systems could, possibly, be the salvation of the cotton industry."

For researchers and policy-makers in pest management—only small amounts of funds are available to pursue promising areas, and relatively little commercial interest is evident *except* when their research leads to a product that can be widely marketed. In the 30 years following World War II, the number of pounds of pesticide used has increased by more than ten-fold. But in that same period, losses to pests remained about the same—about one-third of all crops in the field. Of the 25 most serious agricultural insect pests in California, nearly 75 percent are resistant to one or more insecticide—and 95 percent are either insecticide-aggravated or insecticide-induced pests.

For students—a career in IPM is often a goal but rarely a reality. Many enter agricultural colleges in entomology determined to use their education and training to participate in the development of sound, long-range pest management strategies. "I am a last semester senior at the University of Idaho in Plant Protection (a combination of Entomology, Plant Pathology and Weed Science)," one student recently wrote us. "In attempting to look for job possibilities, my academic advisor has suggested I get a job as a sales repre-

sentative for a chemical company, such as Shell or Gulf & American. I'm positively opposed to working for a chemical company." Excellent courses at land grant universities, including Idaho, train students in biological control techniques. But the private-sector jobs available in IPM upon graduation are scarce.

For consumers and citizens concerned about environmental quality—the dilemma is frustrating, as the divergent interests of society and pesticide makers become ever more obvious. The public has minimal impact upon pesticide use decisions.

For pesticide makers—though sales figures are ever-increasing, pressures from EPA regulations, the costs and red tape of obtaining official clearance for marketing new pesticides, the problems with pollution controls and worker safety at the manufacturing site all add up to major headaches for pesticide manufacturers. The costs of developing a new pesticide are in the range of $15 to $20 million.

Thus, in varying degrees, no one is really satisfied with today's pest management methods, but *no one group by itself* is capable of breaking up the vicious cycle. The solution requires the interaction—and, to some extent, the cooperation—of agricultural extension agents, farmers, researchers, policy-makers, IPM consultants and consumers.

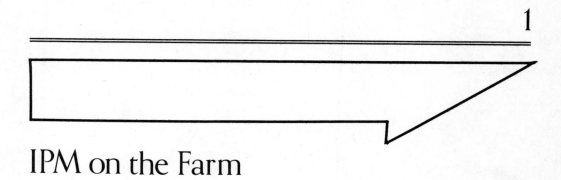

IPM on the Farm

Chapter 2

"Money in the Bank" from Less Pesticides

Apple maggots are trapped by hanging sticky red spheres like this one on trees. Traps also serve to measure the economic significance of apple pest population.

Ed Roberts of Granville, Massachusetts is president of the Massachusetts Fruit Growers Association, and has been a commercial apple grower for more than 30 years. In the summer of 1977, he used 24 sticky red spheres to monitor his 110 acres of apple orchards.

The spheres serve as traps, attracting pests like the apple maggot and capturing them in the sticky surface film. In larger orchards, such as Roberts', the "catch" tells the grower if and when an insect infestation of economic significance is occurring. Pesticides can then be sprayed only when they are needed, instead of on a pre-planned (and blind) schedule. In smaller orchards of less than 150 trees, the traps usually will keep insect damage down to acceptable levels without the use of any pesticides.

"Those sticky spheres saved me two sprayings," orchardist Roberts reported, "and that's money in the bank."

The community in and around Coy,

Arkansas is in one of the best cotton-producing regions in the lower Mississippi River flood plains. Mean annual lint cotton production in the community is about 750 pounds. The bollworm hazard in the area is considered to be one of the highest in the state.

In 1975, the Coy community was chosen to be part of an Integrated Pest Management program conducted by the National Science Foundation (NSF), and the Environmental Protection Agency (EPA), known as the Huffaker Project. After two years, the feasibility of a community-wide IPM approach to cotton insects was unquestionably demonstrated. According to Arkansas entomologists J. R. Phillips and W. F. Nicholson, insecticide inputs for bollworm control were reduced by more than 80 percent. During 1977, this reduction amounted to about $30.00 per acre.

Many similar experiences and studies show that farmers can reduce pesticide use by between one-third and two-thirds, using IPM methods, with negligible profit differential between IPM and conventional pest control. Many farmers, in fact, report increased profits by using *less* pesticides. Yields are maintained with IPM methods and total pest management expenditures are reduced, with more of the outlay going for services of commercial entomologists and scouts. The United States can still be the world's largest food producer without relying so totally upon agricultural chemicals.

At a 1977 symposium on the "Economics of Integrated Pest Management" in San Diego, Darwin C. Hall of the University of California's Department of Agricultural and Resource Economics gave a paper on IPM dollar returns to farmers over a five-year period. In "The Profitability of Integrated Pest Management: Case Studies for Cotton and Citrus in the San Joaquin Valley," Hall stated that "there is no statistically significant difference in profit between supervised control and conventional control for both crops when considering the entire five-year period 1970–1974."

Hall had reported similar findings earlier in the October, 1975 issue of *California Agriculture,* "The Performance of Independent Pest Management Consultants." Co-authored by R. B. Norgaard and P.K. True, this article was based on interviews with 42 cotton and 39 citrus growers. About six percent of the total citrus acreage (6,070 acres) was served by consultants in the San Joaquin Valley in 1970; for cotton, the figure served by consultants was about 20 percent or 126,000 acres. Over the two-year period (1970–1971), IPM user-farmers spent an average of $7.00 less per acre on insecticides than non-users.

In July, 1977, Erik Jansson of Friends of the Earth (FOE) summarized field reports on IPM under the heading, "Is the Farmer Making A Profit from These Programs, and How Much?" Jansson interviewed 37 field managers of IPM programs financed over six years by the USDA Extension Service and by a joint EPA-USDA-NSF program. He concluded that, based upon USDA extension managers, "the farmer is making a substantial profit from IPM programs." (See chart on farmer savings per acre.)

Here are comments from USDA per-

FARMER SAVINGS PER ACRE
FROM THE USE OF INTEGRATED
PEST MANAGEMENT AND BIOLOGICAL
CONTROL OF PESTS

"Dramatic Savings" 1. 20 to 1 return or more to farmer from government investment 2. $100 to $1,000 per acre	"Substantial Savings" $6 to $100 per acre	"Covers Costs of Scouting or Supports Private IPM Firm" $1.50 to $6 per acre
alfalfa, New Jersey (use of parasite)	alfalfa, Ohio alfalfa seed, Oregon animal flies, Calif. animal flies, Nebraska	avocado, California (no treatment)
alfalfa, ten states (use of parasite)	apples, Michigan	alfalfa, Illinois
apples, New York	citrus, Florida	corn, field, Illinois
citrus, Florida (use of parasite)	corn, sweet, N. Jersey	corn, field, Ohio
citrus, California	cotton, Alabama, 1972–4 Arizona, 1976	corn, field, Nebraska
cotton, Texas 1976 (use of early maturing variety)	Arkansas, 1976 Calif., 1976 Louisiana, 1972–4 Mississippi, 1972–4	cotton, Georgia, 1976 New Mexico, 1972–4 (minimum treat.)
pears, California (prediction of fire blight)	S. Carolina, 1972–4 Texas, 1972–4	pecans, Texas (no or minimum treat.)
peaches, S. Carolina (control of nematode)	peanuts, Texas	
soybeans, Maryland & Delaware (use of parasite)	pears, California pecans, Alabama	
	peppers, Delaware	
Weeds	potatoes, Maine	
musk thistle control (use of predator)	potatoes, Delaware sugar beets, Michigan	
northern joint vetch control (use of fungus)	tobacco, North Carolina tomatoes, Florida limabean, Delaware	

sonnel quoted in the FOE report:

Dr. G.P. Dively, Maryland—Soybeans

"We run a program of mass production of parasite wasps for Mexican bean beetle control in soybeans. We are able to carry this entire program at a cost of $50,000 a year, which includes mass rearing, airplane distribution and administration. My estimation is that this program saves the farmers between $1 and $3 million a year.

"Scouting programs back up this program, but the parasite wasp has made it necessary for only five percent of farmers to spray each year."

Dr. B.D. Blair, Ohio—Alfalfa, field corn

"In our area, it is my estimation that in field corn we will generate enough savings to pay for scouting costs. In alfalfa, the return is much more because we are able to save the farmer a new seeding and also generate higher yields by better scouting and assessment of insect pressures."

Dr. G.C. Fisher, Oregon—Alfalfa

"We are in our second season and hope to save two applications per season for the farmer. There is presently a considerable range of spraying schedules, with one farmer spraying 11 times a season. Chemicals for spraying cost between $3.50 and $12.00 per acre, with the cheaper materials being on the restricted list. This averages out in savings from IPM of $7.00 and $24.00 an acre for materials.

"Pollination is even more important to yields than control of pests in alfalfa, since it requires cross pollination. We have been studying the effectiveness of the leaf cutting bee and alkali bee, which are more important than the honey bee. The tripping mechanism of the flower of alfalfa hits the honey bee on the head and discourages them. The leaf cutting bee also cuts holes in the alfalfa and roses, in addition to its critical importance in maintaining yields.

"Our program is being run like that of Idaho, with a field supervisor and six scouts this year.

Attractant-baited traps are placed outside bee shelters in alfalfa fields to lure the checkered flower beetle which is a predator of the widely-used alfalfa pollinator, the leafcutting bee.
(*Agricultural Research, USDA*)

The two alternatives to IPM are calendar spraying and reliance on the advice of seed and chemical salesmen. It is not known whether this advice is reliable or whether it is overprescribed. But the critical lack of these salesmen is their inability to identify and quantify the beneficial control factors in the field."

County agents Leonard Cobb and Mike Linker in Marianna, Florida are scouting 9,000 acres of peanuts and soybeans, coordinating the program with private firms who are paid on a contract basis at $2.50 per acre. "Farmers are saving money and are pleased with the program," says Cobb, "but I've never worked so hard in my life." He also points out that county agents still need to get involved because the farmers are "not sold on it yet."

In Homestead, Florida, Dr. Kenneth Pohronezny of the University of Florida is training scouts to work in local tomato fields. The three persons trained thus far took an intensive two-month course and are now continuing their learning in the fields. They're paid $7,750 per year and are employed full time. Two are college graduates, and the third has had farming experience.

Each scout can cover approximately 100 acres of tomato fields per week. For soybeans, a scout can survey about 1,500 acres per week.

At the Pennsylvania State University Research Center, Dr. Zane Smilowitz and associates have been using integrated methods on potatoes for two years. With no insecticides and lower-than-normal quantities of soil systemic insecticide, yields were equal to those of plots with standard amounts.

Extension entomologists at the University of Maryland have been very active in initiating Integrated Pest Management programs. Their first IPM pilot project involved sweet corn and lima bean growers on the Delmarva peninsula. Originally funded by a federal grant, this program is entirely self-sustaining, paid for by the commercial canners who became convinced of the value of IPM on their acreage contracted to individual growers. Dr. Galen P. Dively, extension entomologist at the University of Maryland (previously quoted), reports that the IPM program is rapidly becoming a truly inter-disciplinary effort. Entomologists are no longer fighting the battle alone. They are being joined by extension weed control specialists and other agronomists.

With the 1978 cropping season, field scouts are being hired in areas where sufficient numbers of farmers have expressed an interest in participating. The cost to farmers will be $2.00 per acre, with a minimum of 50 acres per grower required. About $5,000 will be needed to pay for one scout per season. A minimum of 2,500 acres in one locality will be required to make hiring a scout worthwhile in that area.

Dr. Dively is convinced that "farmers who participate can save money—and sleep better at night—compared with the too-common current method of hit-or-miss application of agrichemicals for 'insurance' purposes."

The USDA estimates that approximately seven million acres of cropland in the United States were under some form of IPM treatment in 1977. The most significant use has been on cotton.

"Good, well managed Integrated Pest Management systems could, possibly, be *the salvation of the cotton industry*," says Research Entomologist James W. Smith of the USDA's Bioenvironmental Insect

Control Laboratory in Stoneville, Mississippi. The reason: many producers in the Deep South cannot control the tobacco budworm despite repeated applications of insecticides costing upwards of $100 per acre.

According to David Pimentel of Cornell University, cotton receives nearly 50 percent of all insecticides used in agriculture; the largest percentage of acreage treated is in the southeastern and delta states.

In the past four years, the New York State Tree Fruit Pest Management Program, managed by Jim Tette, has evolved from one which was federally-supported and available to a few growers, to one which is now grower-supported and includes nearly 15 percent of the tree fruit acreage in New York State. This crop protection program (known as the "Farmer Advisor" program) on all tree fruit pests begins with pre-season grower conferences and strategy meetings, followed by in-season grower advice, recommendations and consultations.

Explains Dr. Tette: "This method has been available to all growers in Wayne County, the largest fruit growing county in New York. Participation in this program has increased from 16 growers in 1975 to 34 growers in 1976 and now includes 12 percent of the total fruit acreage in the county. Grower support increased from $2,400 in 1975 to $28,000 in 1976. The program is carried out by farm advisors who have been trained in insect, disease and mite management by research and extension personnel. A pest management assistant provides the farm advisors with routine information on weather, traps, leaf brushings and laboratory tests."

Growers in the "Farm Advisor" program have been able to eliminate two to three insecticide sprays through careful use of a yellow sticky bait trap for the apple maggot fly in combination with other orchard monitoring.

Extension personnel use monitoring techniques to measure the impact of pesticides on secondary pests, target species, natural enemies and the environment.
(*California Agriculture*)

In commenting upon the results, Dr. Tette makes these observations:

"Some benefits might best be seen by comparing the records of pesticide use of a grower who left the IPM program and two who joined the program in 1976. Records of pesticide use by grower A, who left the IPM program after having been a member for three years, show that his insecticide use nearly doubled after leaving the program and that his cost per acre went from $87.27 to $123.84 in 1976. Although the percentage of clean fruit went from 81 percent to 99 percent, a substantial increase, several important facts about grower A must be understood. In 1975, grower A, despite IPM recommendations to the contrary, chose to use a fungicide for which apple scab had developed a tolerance. Virtually all of the injury in the 19 percent category was caused by apple scab. Insect injury did not change in either year. Examination of records of two growers not in the IPM program in 1975, but members in 1976, show that grower B decreased his insecticide use by more than half and his cost per acre by nearly half while maintaining fruit quality. On the other hand, grower C decreased his miticide and insecticide use and pesticide costs only slightly while maintaining fruit quality.

"Preliminary evaluation of the 1976 IPM efforts also indicates that growers benefitted from improved pest control, reduced production expenses (pesticide savings) and improved management capabilities. In the latter case, many growers benefitted from seemingly insignificant services such as accurate orchard mapping, including acreage measurements and block numbering. Many large growers benefitted from data organization, especially from having spray records on each block throughout the season. Smaller growers benefitted from the individual and thorough attention not often available to them in the past."

Applying Research for Massachusetts Fruit Growers

Dr. Ronald J. Prokopy of the University of Massachusetts' Department of Entomology has recently developed an Integrated Pest Management research project for apple pests. It builds on knowledge gained from past research and has received considerable support from the Massachusetts Fruit Growers Association. Observes Prokopy:

"The recently established arthropod pest management research program on apples in Massachusetts bodes well for implementation of an apple pest management program that will result in real reduction of insecticide and miticide usage."

Here are the new approaches:

1. Use of traps and other appropriate techniques for monitoring populations or incidence of principal key and secondary pests in grower orchards. The traps, Prokopy and colleagues found, monitor plant bugs, sawflies and apple maggots. The traps for monitoring sawflies, for ex-

ample, are 15-by 20-cm rectangles painted with titanium oxide white to imitate opened apple blossoms. For apple maggots, the 8-cm sphere is painted dark red to mimic nearly ripe apples. ("The spheres have proven highly effective in directly capturing out and controlling the apple maggot in small orchards, but require too much labor as a direct maggot control measure in orchards greater than two hectares," says Prokopy. The effectiveness of these spheres was noted in the opening paragraphs of this chapter.)

2. Use of such pest monitoring information in conjunction with economic threshold (action) levels as an index of when some form of treatment against a pest is needed to avoid economic injury to the crop.

3. In the first years of the project, use of pesticides as the principal method of treatment against pests. The only types of pesticides (as well as growth regulators and foliar fertilizers) employed will be those which are comparatively economical for the grower to purchase, effective against the pest in question and have minimal effects (as revealed by present and future research and adaptive studies) on non-target organisms such as predators of spider mites and apple aphids.

4. In selected orchards, combination of the above methods with an alternate middle spray program.

5. Cost-benefit analyses (including all fixed and variable costs of production, yields of marketable produce and receipts for produce) comparing the every-middle pest management spray program, the alternate-middle pest management spray program, and the traditional every-middle, preventive-type spray program.

6. Monitoring for incidence of apple scab resistance to fungicides (principally Benlate and Cyprex).

7. If available for commercial use in the last years of the project, implementation of pheromones as a non-pesticidal means of controlling codling moth, leafrollers and apple maggot.

Prokopy sees these results from the research program:

"If the above objectives can be realized, there should be at least a 30 percent reduction (possibly 50 to 60 percent reduction, depending upon efficacy of alternate middle program) in pesticide usage in those orchards using the pest management program. This would result in substantial economic savings to the growers, greater abundance of natural enemies, a less contaminated fruit product and environment, and a slower rate of development of pest resistance to orchard pesticides."

The research procedure would involve entomologists, economists, pomologists, plant pathologists and scouts. (See organizational chart.)

ORGANIZATIONAL CHART

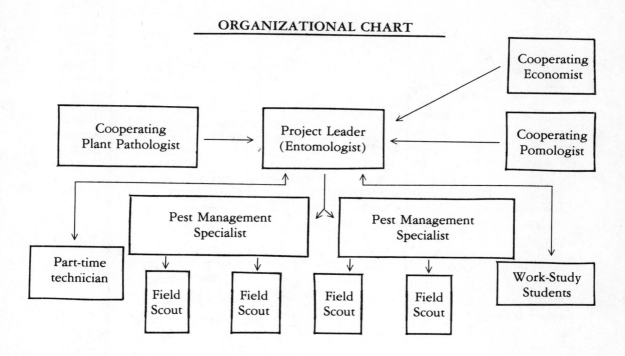

Energy Savings

Since most insecticides are petro-chemicals, pest controls using alternative strategies clearly conserve oil reserves. In a USDA publication series, entitled "A Guide to Energy Savings," the report *For The Vegetable Producer* includes the following:

With an integrated pest control program, growers can usually reduce the number of required spray applications over the long run. Assume the number of applications required is reduced by one on 10 acres of vegetables:

Calculations:
10 acres × 2 gal/acre = 20 gal
Value of fuel saved at 45¢/gal, 20 gal = $9

Diesel Savings at Various Fuel Prices

Cents/gal	35¢	45¢	55¢
Savings/year	$7	$9	$11

Fuel Savings $9

Labor savings × 1 hour/acre @ $3/hour = $30

Net savings $39

Experiences of Organic Farmers

Any discussion of farmers' successful experiences in using less pesticides should definitely refer to the practices of organic farmers. Organic farming methods are usually defined in negative and positive terms: negative—avoiding artificial fertilizers like anhydrous ammonia, superphosphate and chemical pesticides; positive—applying organic materials (manure, compost, etc.), cropping systems that include green manures and soil-building rotations.

Case studies have appeared in the pages of *Organic Gardening and Farming* magazine since it was first published in 1942, and more recently in such publications as *Countryside, Tilth, Maine Organic Farmers, Midwest Organic Producers, Harrowsmith* and others, which show that farming is even profitable without any use of pesticides.

An overview of organic farming methods was presented in *Organic Farming; Yesterday's and Tomorrow's Agriculture,* edited by Ray Wolf (published by Rodale Press). Here, for example, is how one Kentucky organic farmer, Jim Foote, approaches weed control on his 300-plus acres.

"Our weed control effort was aimed at getting all of the weeds we could without the use of chemicals. Notice I didn't say *all* of the weeds like some of the experts stress. But I'm more than satisfied with our weed control this year compared to the best chemical farmers around. Corn and soybean ground was plowed in the early spring with a moldboard plow and planted as early as possible. I considered delaying planting until later in order to destroy more weeds with the disc harrow, but decided to plant early instead. The soybeans were cultivated once with a rotary hoe and twice with a spring-shovel cultivator. Part of the corn was rotary-hoed, about half was not. All of the corn was cultivated once, most of it twice. I couldn't tell much difference as long as it got one good *early* cultivation. Weed control in the soybeans was excellent—as good as I'd hope to see. Weed control in the corn ranged from fair to good. We had no crop that suffered from weeds. I found that one trip over with the rotary hoe and two trips with the cultivator gave me excellent control of weeds, provided the first cultivation was timely.

"Weeds were controlled in the pastures by clipping. Part of the pasture was clipped with a rotary and part with a sickle-bar mower. I lean toward the rotary mower because of the way it shatters the residue, leaves a taller stubble for cattle to graze and requires less maintenance."

John Adams raises beef and grain organically on his 400-acre farm in Kansas. His rotation is soybean, corn, wheat and clover for green manure and nitrogen fixation. "We've got sweet clover coming up all over the place like a noxious weed," says Adams with a smile. "And I think that's awfully nice." In addition to its fertility value, sweet clover controls troublesome bindweed, Adams has found. He does have problems with cocklebur in his hayfields, though he gets some control from periodic mowings of the hay.

The Center for Rural Affairs in Walthill, Nebraska surveyed 140 Nebraska commercial farmers who used no inorganic chemicals on a substantial portion of their farms. A brief summary relating to how their methods affected weed and insect problems follows:

The typical operation surveyed consists of corn, oats, soybeans and alfalfa grown in a strict rotation program. The purpose is to keep soil fertile and productive without applying any oil-based synthetic fertilizers. To overcome weed problems, farmers plant later in the year. This has the two-fold benefit of allowing them time to disc under weeds twice before sowing, and gives the crops a chance to shoot

Crop rotation helps to control western rootworms which can severely reduce corn yields as shown by the two ears on the right, damaged by silk feeding adults.
(Agricultural Research, USDA)

up rapidly in the warm soil. During the growing season, says Del Akerlund of Valley, Nebraska, weeds are "taken care of through meticulous types of cultivation, rotation and some hand care."

In addition, pointed out many Nebraska organic farmers, rotation keeps soils looser and mellower (hence less weed troubles), and also keeps most pests from proliferating. As the Bergman brothers of Elgin state: "When you rotate crops, you don't have to worry about rootworms, corn borers and many others."

Concludes the Center for Rural Affairs account of organic farmers:

"Because of the farmers' feeling that they are working in tune with nature, accepting the rhythms of nature, they appear to feel a contentedness, a 'rightness' about their way of farming. They switch from chemical farming not for reasons of financial gain, but because they are disturbed by the lack of wildlife in their fields, or their ever-harder soil or the residues of toxic chemicals they fear ingesting; they feel there is something 'not right' with a way of farming that causes these problems. The decision to go with natural farming is for many a decision about the quality of life they wish to lead."

Contrary to so many of the commercials and advertisements for herbicides and insecticides, not every United States farmer is awe-struck by the "magical protective powers" of these products. More and more farmers are voicing their growing dissatisfaction. However, few are able to develop a plan of action for reduced pesticide use. The reasons most farmers continue to rely on pesticides need to be explored as do the motivations of those few farmers who have switched entirely to biological methods.

Chapter 3

Motivations: Why Most Farmers Use Pesticides . . . And Why Some Do Not

If Integrated Pest Management is to become genuinely significant on the American farm scene, then obviously farmers must develop confidence in IPM and the firms which sell IPM services. How will that happen? On what criteria do farmers decide to use, or not use, certain pesticides? And what changes need to take place for IPM methods to become more widely used?

Unfortunately, though modest amounts of research funds have yielded significant data on specific methods and benefits, scarcely any projects have addressed the above questions. In fact, we've only come across two studies, both of midwestern Corn Belt farmers, which offer relevant data. One is "Farmers' Pesticide Use Decisions and Attitudes on Alternate Crop Protection Methods"; the other is "Motivations and Practices of Organic Farmers."

Farmers' Pesticide Use Decisions

In 1974, the EPA's Office of Pesticide Programs and the Council on Environmental Quality jointly financed this study. Dr. Rosmarie von Rumker of RvR Consultants served as the principal investigator of the 297 farmers in Iowa and Illinois who supplied comprehensive answers about their farming, crop protection practices and information services.

Farmers in the Midwest, she found, receive information on pesticides primarily from pesticide sellers, labels and other farmers. Pesticide industry representatives and sellers outnumber extension personnel by such wide margins that extension messages (about IPM) do not reach a significant number of growers directly.

"Pesticide information flow patterns developed during a period when the objectives of the pesticide trade were essentially in harmony with those of federal and state agencies," wrote Dr. von Rumker. "In the past, pesticide development and increased use of pesticides generally meant progress to both groups. *Messages recommending reduced or no use of pesticides do not flow well through this system.*" (Emphasis added.)

The farmers who were interviewed chose pesticides because, based upon the available information, the chemicals were more profitable than alternatives. However, the study revealed that farmers are generally unaware of how current crop protection decisions may entail hidden future costs, and that *no effort is being made to bring it to their attention.* Two important types of side effects, resulting from pesticide use, are usually neglected by farmers. Dr. von Rumker explains:

"First, surprisingly little information is available on the interactions between different pesticides, and between pesticides and all other elements of crop production systems. For instance, chemical herbicides, insecticides, fungicides and fertilizers are often applied to the same land year after year. Most of these chemicals remain in the upper one to three inches of the soil. Little is known about their routes and rates of degradation under field conditions, on their individual and collective effects on the soil microflora and microfauna and on the long-term fertility of the topsoil.

"Second, the past history of the use of pesticides teaches that the usefulness of most or all of these chemicals is gradually eroded by the development of resistance in the target pest(s), the selection of pest populations that are increasingly more difficult to control, the destruction of beneficial organisms and other factors.

. . . Information on these (and other) side effects that are currently largely disregarded would provide farmers and those advising them with a better basis for making ecologically and economically sound crop protection decisions."

All farmers interviewed (58 in Iowa, 239 in Illinois) raised corn, and more than 95 percent also raised soybeans on an average acreage of 550. According to USDA estimates, the rapid rise of pesticides used by farmers in this region was for corn herbicides (more than half of the total pesticides used), followed by soybean herbicides and corn insecticides. Table 1 lists responses to a variety of questions concerning pesticide use. While almost all respondents considered herbicides an essential crop protection practice, only 53 percent (Iowa) to 70 percent (Illinois) used corn insecticides in 1973. And only about one-third of all corn growers believed that corn acres need insecticide treatments each year.

Corn growers were also asked if they would change their corn herbicide use practices if the cost of herbicides was twice as high. Those respondents who said yes were asked what they would do. The answers to this question indicate what possibilities growers see to reduce use of herbicides on corn.

	Iowa	Illinois
	% Respondents	
Cultivate more	50	43
Reduce corn acres	36	6
Use lower rate of herbicide	14	13
Band herbicide rather than broadcast	14	8
Treat fewer acres with herbicide	5	26

Table 1: **Iowa and Illinois Farmers' Crop Protection Practices, 1973.**

% Respondents
Iowa   Illinois

Corn Herbicides.

Iowa	Illinois	
100	99	——used corn herbicides in 1973.
70	42	——changed products at least once since 1970.
97	94	——considered herbicide efficacy satisfactory or better in 1972 and 1973.
88	82	——believe all of their corn acres need herbicides each year.
62	56	——would not change herbicide use if cost would double.
97	97	——would continue to grow corn if no herbicides were available.

Corn Insecticides.

Iowa	Illinois	
53**	70**	——used corn insecticides in 1973.
52*	43*	——changed products at least once since 1970.
100*	94*	——considered insecticide efficacy satisfactory or better in 1972 and 1973.
55*	55*	——believe all of their corn acres need insecticides each year.
29**	39**	
77*	61*	——would not change insecticide use if cost would double.
97*	94*	——would continue to grow corn if no insecticides were available.

Soybean Herbicides.

Iowa	Illinois	
98	98	——used soybean herbicides in 1973.
58	45	——changed products at least once since 1970.
92	85	——considered herbicide efficacy satisfactory or better in 1972 and 1973.
96	86	——believe all of their soybean acres need herbicides each year.
72	67	——would not change herbicide use if cost would double.
76	79	——would continue to grow soybeans if no herbicides were available.

* % of corn insecticide users
** % of corn growers

Less than five percent mentioned changing crop rotations, planting dates or corn variety.

Dr. von Rumker noted that many authors have explored the process of how farmers adopt new ideas, technologies and farm practices, but that only one specifically dealt with what farmers know about pesticides. In 1965, Beal et al. conducted personal interviews of 229 farmers who farmed at least 70 acres and personally made the major management decisions. Respondents were asked to indicate which of 29 specified sources of information on pesticides they were then using. Responses were as follows:

	% of Respondents
Farm magazines and farm papers	94.3
Pesticide label	90.4
Other farmers in the community	67.7
Local agricultural chemical dealers	60.7
Radio	48.9
County extension personnel	47.6
Veterinarian	43.7
Television	42.4
Agricultural chemical company publications	41.5
Newspapers	35.8
Agricultural extension publications	34.9
Meetings sponsored by local agricultural chemical dealers	25.3
Family members	25.3
Meetings sponsored by ag. chemical companies	24.9
USDA publications	20.5
Agricultural chemical manufacturers' or wholesalers' salesmen	17.9
Meetings sponsored by the State University and Extension Service	16.6
Vocational agriculture teacher	14.4
State University specialists	14.0
Iowa Farm Science publication	10.0

Additional questions were asked to determine the most useful source of information in regard to specific aspects of the use of pesticides. The pesticide trade, especially local dealers, and farm magazines and farm papers were rated most useful regarding which chemical to use for a particular purpose. Pesticide labels were rated most useful regarding methods and rates of application, handling and safety precautions, and hazards and possible harmful consequences of misuse.

Motivations and Practices of Organic Farmers

One major step in the development of "least-is-best" pesticide strategies would clearly seem to involve careful studies of the *motivations,* as well as the techniques, used by organic farmers. Depending upon the region of the country and specific crops being grown or animals raised, such techniques would include rotations, cropping systems, conservation practices, management methods, marketing methods and economic results. An understanding of attitudes is most significant to researchers and policy-makers concerned with pesticide decision-making of farmers.

Several studies have shown that yields on organic farms are satisfactory in terms of economics and production compared to neighboring farms where chemicals are applied at conventional rates. These studies have been conducted and reported by the Center for the Biology of Natural Systems, Washington University, St. Louis, Missouri—*Organic and Conventional Crop Production in the Corn Belt;* by Frank P. Eggert, Plant and Soil Sciences Department, University of Maine—*The Effect of Several Soil Management Systems on Soil Parameters and*

on the Productivity of Vegetable Crops; by Ray Wolf, Organic Gardening and Farming magazine—*Organic Farming: Yesterday's and Tomorrow's Agriculture;* and by H. Vogtmann, Research Institute of Biological Husbandry, Oberwil, Switzerland—*Effects of Agricultural Practices on Soil and Plant Quality.*

Studies examining the pest management practices of farmers who now are not relying on chemicals should not and need not be carried on as if they were part of a war between organic fanatics on the one hand and rabid pesticiders on the other. The simple fact is that the methods used by organic farmers and their results constitute *valid models* as better pest management systems for the future are sought.

In 1977, the National Science Foundation supported a research study of "Motivations and Practices of Organic Farmers." Sarah Wernick of Washington University's Department of Sociology and William Lockeretz of the Center for the Biology of Natural Systems conducted the survey of 302 commercial-scale Corn Belt organic farms.

(The excerpts below are taken from the report, authored by Sarah Wernick and William Lockeretz of Washington University, St. Louis. The complete text appears in the November–December, 1977 issue of *Compost Science,* published in Emmaus, Pennsylvania.)

The fact that individual large-scale organic farms can be economically competitive challenges the belief that present-day organic farmers can perform no better than farmers of 40 years ago, and that they have no relevance to modern-day agriculture. Organic farmers therefore present an interesting case of unconventional management. They have rejected modern fer-

tilization and pest control methods, which are generally considered indispensable to modern agriculture. Yet at the same time they try—with reasonable success, apparently—to attain a level of production and profitability that enables them to survive in competition with farmers who are using all the modern farm inputs. But the differences between organic and conventional

management are of more than just academic interest. Environmental and energy factors may cause more farmers to make greater use of some techniques that organic farmers (as well as many conventional farmers) are already using, such as rotating forage legumes like alfalfa with row crops, and recycling of livestock manures to cropland. In a sense, then, these "old-fashioned" techniques are becoming fashionably modern. It is therefore interesting to consider why organic farmers—who now are making considerable use of certain methods that many other farmers eventually might adopt to a more limited degree—have chosen this system, and how this decision was related to other aspects of their approach to farming. To explore this topic, we have made a survey of operators of Midwestern commercial-scale organic farms.

Although our criteria to define organic farming were all negative (i.e.,

materials that a farmer may not use if we are to consider him organic), we know from other studies that organic farmers, as defined here, tend to use certain positive practices to a greater degree than do conventional farmers. For example, they have a greater proportion of their cropland in rotation meadow, and less in row crops, compared to conventional farms. Manuring practices are similar on organic and conventional crop-livestock farms, but a larger proportion of organic farms have livestock.

A total of 302 farmers met the following criteria for inclusion in this study: they had to raise field crops primarily, farm at least 100 acres and use organic methods on more than half of the land they managed.

Respondents managed a median of 244 acres. Five-sixths managed all their land organically. Twenty-three of the respondents had always farmed organically. For the remainder, the median year of conversion to organic methods was 1970. About three-fifths of the farmers were between 40 and 60 years of age, with approximately equal numbers above and below that range.

The questionnaire was designed to answer the following:

1 What were the motives and decision-making processes of the farmers who had converted to organic management?

2 What inputs did they use and which did they avoid?

3 Apart from their non-use of certain chemicals, to what extent were their equipment and practices modern?

4 What contact did they have with other farmers and with agricultural institutions, both conventional and organic?

Findings

The Decision to Farm Organically

One hundred fifty-one respondents had converted from conventional to organic management. We questioned them about their motives for this change, the advisors they had consulted and the time frame in which the decision had been made. All respondents, including the 23 who had always been organic, were asked to indicate advantages and disadvantages of organic farming.

Motives for conversion to organic methods

Motives for changing to organic management varied considerably, although the reasons given generally fell into three categories: specific problems with conventional farming, ideological misgivings about conventional farming and contact with an individual or organization suggesting organic farming as an alternative. Table 1 summarizes the answers most frequently given.

Problems with conventional farming were mentioned by three-quarters of the farmers who formerly used conventional methods. About a third had been concerned about livestock health. Nearly as many were persuaded to convert by deteriorating soil quality. Several respondents wrote, "The soil was dead." A third of the farmers expressed some ideological criticism of chemicals; less than one-tenth cited environmental or religious objections.

Contact with a proponent of organic farming motivated half the respondents to consider a switch, with about one-third mentioning organic farming product salesmen. As will be discussed later, these salesmen play an important role in supporting organic farming. For the most part, the farmers were predisposed to consider organic farming before contact with salesmen.

The typical scenario included all three elements. A generally critical attitude towards chemical inputs was galvanized by some specific problem with their performance. Subsequent contact with organic farming enthusiasts further convinced the farmer of these drawbacks and suggested an alternative.

Advisors on adopting organic methods

Once seriously considering organic farming, all those who formerly farmed conventionally sought some outside advice and assistance. As Table 2 shows, the organic farming product salesman was the most significant change agent. In contrast, a few of these farmers had gone to more conventional sources of information: only five farmers had consulted the Soil Conservation Service (SCS) and just two talked with Extension agents.

Timing of conversion to organic methods

Of organic farmers who formerly used chemicals, four-fifths converted to organic methods within the past decade, with the median date being 1970. Virtually all of these farmers began using organic methods the same year that they first considered the change, and most converted all their land by the first or second year.

Eighteen percent indicated that their

TABLE 1. Reasons for Converting to Organic Farming.

	% of Respondents Mentioning Reason
Specific Problems or Concerns	
Livestock health	32
Soil problems	30
Cost of chemicals	25
Chemicals not effective	23
Human health	22
At least one specific problem or concern	75
Ideological Concerns	
Dislike chemicals	24
Environmental concern	8
Religious concern	5
At least one ideological concern	34
Contact with Proponents	
Dealer for organic fertilizers and soil amendments	30
Farmer	14
Organization	8
At least one contact	48

TABLE 2. Sources of Advice on Decision to Adopt Organic Methods.

	% Who Consulted Indicated Source
Salesman for organic farming products	84
Organic farmer	72
Books or articles	61
Soil Conservation Service	3
Extension Service	1

net income dropped temporarily after they began farming organically. Sixty percent reported no drop, while the remaining 22 percent were uncertain, either because their records were incomplete, or because losses appeared to have been caused by drought or other adverse farming conditions unrelated to the conversion.

Perceived advantages and disadvantages of organic farming

The farmers studied were highly committed to organic farming. Of the 32 who had not converted all their land, 16 planned to do so, while ten of the remaining 16 could not convert rented land because of the owners' objections. Obviously, a decision to become organic can be readily reversed. Twelve percent had considered (then decided against) returning to conventional methods, most frequently because of low yields or weed problems.

The fact that most of these farmers have never even considered giving up organic methods and that most are or will be managing all of their land organically, indicates a strong commitment to this way of

TABLE 3. **Perceived Advantages of Organic Farming.**

	% Indicating That This Was One of the Three Most Important Advantages
Healthier for the farmer and his family	54
Healthier for livestock	48
More wholesome, in harmony with nature	28
Better for the soil	26
Better for the environment	25
Closer to the way of farming in the Bible	20
Lower production costs	18
Higher net income	12
Higher quality product	11
Tillage easier	10
Yields suffer less under adverse conditions	9
Consumes less energy	8
Less dependence on outside suppliers	7
Fewer insects	6
Crop dries better	4
Fewer weeds	4
Total hours of labor less	1
Yields higher	1
Organic products command a premium price	1
Work more evenly spread through the crop year	0

farming. Therefore, it is not surprising that these respondents reported considerably more advantages than disadvantages to organic farming. When presented with lists of both, a mean of 13.3 advantages were checked, but only 2.8 disadvantages. Tables 3 and 4 show the advantages and disadvantages which they considered most important.

The safety of organic farming, for both livestock and farmer, is its most significant advantage in the opinion of these respondents. Of the four most frequently mentioned disadvantages, only one—weed problems—is agricultural. The other three problems—finding suitable markets and up-to-date information and the negative at-titudes of others—reflect the social context of organic farming. These difficulties suggest some of the reasons they have turned to organic farming institutions.

Inputs Used

Although small-scale certification programs do exist, there is no universally recognized standard of organic farming. Therefore, we sought to determine the degree of consensus about the acceptability of certain inputs in organic crop and livestock operations.

We asked respondents whether they had used, might use or would not use certain inputs on their organic cropland. The results are presented in Table 5.

TABLE 4. **Perceived Disadvantages of Organic Farming.**

	% Indicating That This Was One of the Three Most Important Disadvantages
Difficult to find a market for organic products	22
Weed problems worse	16
Fewer up-to-date sources of information	16
People look down on organic farmers	14
Difficult to get enough manure	10
Greater expertise needed	9
More labor required	8
Can't have all land in cash crops	5
Harder to get loans	5
Lower yields	4
Lower profits	2
Pests worse	2
Organic farming requires livestock	1
No time to take vacation	0
No disadvantages	13

TABLE 5. Acceptability of Selected Inputs for Organic Cropland.

Substance	% of Respondents Who Would Never Use Indicated Substance*
Anhydrous ammonia	93
Insecticides	85
Herbicides	75
Urea	70
Super phosphate	60
Grain silage preservative	45
Sewage sludge	37
Potash	27
Rock phosphate	21
Trace minerals	8

*It is possible that some farmers did not understand the question and included inputs on non-organic land they manage, or referred to their practices before conversion to organic methods. When it was clear from marginal notes that such misunderstandings had occurred, the respondent was eliminated from the tally unless the answer could be extracted from the comments.

Chemical fertilizers, insecticides and herbicides—substances used by the vast majority of conventional farmers—are, of course, unacceptable to most organic farmers we surveyed. Farmers who had more contact with organic farming institutions—dealers, organizations or publications—more frequently indicated that they would never use these inputs.

Modern Equipment and Practices

This study focused on farmers who had decided, for a variety of reasons, to reject certain modern farming inputs. It is of interest to see if this decision was general or selective.

While there was considerable varia-

tion in this regard, the majority of the farmers had adopted at least some modern equipment and practices.

Relationship with Farmers and Agricultural Institutions

Relations with conventional farmers

Two-fifths of the farmers reported as a disadvantage of organic farming that others tend to look down upon this management system. As mentioned earlier, 14 percent considered this one of the three leading disadvantages. Nevertheless, over four-fifths said that they did discuss organic methods with conventional neighbors. According to the respondents, most neighbors were indifferent or skeptically interested in organic methods; only 14 percent said that most of their neighbors were hostile.

Apart from ideology, relations between organic and conventional farmers can be strained by problems arising from differences in management. On the one hand, organic farmers can be plagued by drifting chemical sprays or chemicals in runoff water. On the other hand, organic farmers may be viewed as sources of weed or insect pests. We asked respondents to tell us (in a necessarily one-sided report) if they had experienced such difficulties. Nearly half (46 percent) said that they had had some problem from a conventional neighbor, usually drifting spray. Those who marketed crops through organic channels were more likely to report this. Only five percent reported complaints against themselves, mostly about weeds.

Contact with conventional farming institutions

Almost half (47 percent) of the respondents had had some formal conventional agricultural training, usually through high school courses (33 percent), GI farm training (16 percent) or college courses (nine percent). Moreover, they reported about as much contact with conventional as with organic farming institutions: these farmers are not isolated as individuals or as a group. Table 6 gives data on three different kinds of contact with farming institutions: magazines, organizations and meetings. The more institutional contact a farmer had, the more likely he was to be using modern equipment and practices. A few of the farmers stated in marginal comments that they read conventional magazines for an unconventional reason: "to see what the enemy is up to."

Relationship with organic farming institutions

Although contact with organic agricultural institutions did not preclude contact with conventional farming institutions, it clearly played an important role in supporting these farmers' activities. Table 6 has already shown the degree of contact that respondents had with other organic farmers through magazines, organizations and meetings. Four-fifths read at least one organic farming magazine; half belonged to at least one organic farming organization; and three-fourths had attended at

TABLE 6. **Contact with Agricultural Institutions.**

Form of Contact	Organically Oriented	Conventionally Oriented
Mean number of magazines read	1.7	2.4
Mean number of memberships in organizations	.8	.8
Mean number of meetings attended in 1976	1.1	1.0

least one meeting related to organic farming in 1976.

We have already seen that commercial dealers for organic products, such as fertilizers and soil amendments, are a particularly significant change agent. After a farmer adopts organic methods, a dealer plays an important role as a source of inputs, advice and encouragement. Just over half of the respondents had attended a farmer-dealer meeting sponsored by an organic product, and about one-third belonged to a particular organization of farmers who use one of these products. As Table 7 shows, nine-tenths of the farmers were using some purchased organic materials in their farming. Many companies supply not only their product, but also a complete management program, including suggested crop rotations, tillage methods, etc. About half of the respondents were following such a program without modification.

There was a significant relationship between the year a farmer adopted organic methods and his involvement with a commercial product: those who changed recently are more likely to follow a complete commercial program. Farmers who have always farmed organically are more likely to use no product at all, or simply to use a product without the rest of the program. Moreover, these farmers had significantly fewer contacts of any kind with organic farming institutions compared to farmers who converted.

It was stated earlier that the disadvantage of organic farming most cited by respondents was difficulty in finding an organic market for their products. Indeed, only a minority (33 percent) use special organic marketing channels at all, and just a third of these (12 percent of the sample) are able to sell at least half of their products this way. Of those who use special channels at all, five-sixths receive a premium price.

Because organic products command a premium on the retail market, it is sometimes assumed that organic farmers routinely receive higher-than-market prices for their goods, which is considered essential for them to survive economically. Actually, only ten percent of the farmers were receiving premium prices on at least half of their crops or half of their livestock, and 18 percent were getting premiums on occasional sales (usually to individuals). It was clear that many more farmers would

TABLE 7. **Use of Commercial Organic Products and Programs.**

Use of Commercial Organic Product		% of Respondents
No use		11
Some use		89
Product only	19	
Product and modified version of program	19	
Product and full program	51	

have used special channels had they been available.

Farmers who charged a premium stated that the higher price was not the result of higher production costs or extra expenses in marketing the product, such as special bagging. Rather, about four-fifths of these farmers said that the higher price was charged because of the superior quality of the product, and because customers were willing to pay more for organic merchandise. Given the farmers' readiness to charge higher prices, and the customers' willingness to pay them, it is reasonable to expect further development of special marketing channels, possibly through existing organic organizations. Although few of these organizations presently offer such a service directly, contact with organic farming institutions was significantly associated with a farmer's ability to locate an organic market for his output.

Organic farming institutions also compensate for two other difficulties reported by the respondents: lack of up-to-date information and the negative attitudes of others. These institutions offer technical advice and interpersonal support to the farmer whose use of organic methods precludes his seeking such help elsewhere.

Discussion

Many people apparently have an image of organic farmers that is very different from what emerges from this study, judging by the correspondence we have received and the questions we have been asked in connection with our larger study of organic farming. It is sometimes assumed that organic farmers usually use animal power for tillage, that they mainly raise fruits, vegetables and specialty crops specifically for the organic or natural foods market and that they tend to operate marginal, subsistence enterprises. It is also thought by some people that organic farmers represent a "back-to-the-land" movement that is an alternative to prevailing life styles and economic institutions.

While some organic farmers do fit this image, the overall picture of those we surveyed is very different. They raise field crops and livestock on a commercial scale, using mechanized methods. Most of their output is sold through conventional channels. Most farmed by conventional methods before changing to organic methods. Besides being involved with organic farming institutions, they tend also to maintain contact with conventional agricultural institutions, such as magazines and organizations. Their reasons for choosing organic methods generally were practical, that is, related to specific problems with conventional farming, such as possible adverse health effects of pesticides. Ideological or philosophical considerations also played a role, but to a lesser degree. In summary, it appears that for the type of organic farmer covered by our survey, choosing to use organic methods does not necessarily imply an overall approach to farming that sets organic farmers apart in a fundamental way from the conventional farmers who comprise most of Midwestern agriculture.

References

R. Klepper, W. Lockeretz, B. Commoner, M. Gertler, S. Fast, D. O'Leary, and R. Blobaum: "Economic Performance and Energy Intensiveness on Organic and Conventional Farms in the Corn Belt: A Preliminary Comparison." *Amer. J. Agr. Econ. 59*, 2 (1977).

W. Lockeretz, R. Klepper, B. Commoner, M. Gertler, S. Fast, D. O'Leary, and R. Blobaum: "Economc and Energy Comparison of Crop Production on Organic and Conventional Corn Belt Farms," in *Agriculture and Energy,* proceedings of a conference held in St. Louis, Mo., June, 1976, W. Lockeretz, Ed. (Academic Press, New York, 1977), pp. 85–101.

Quoted in *Yearbook of Spoken Opinion: What They Said in 1971,* p. 123.

U.S. Dept. of Commerce, *1969 Census of Agriculture,* Section 1 of individual state volumes.

Helping Farmers to Use Less Pesticides More Successfully

The findings of these two studies suggest that certain changes are necessary in order to genuinely assist farmers in using minimum amounts of pesticides for maximum advantage.

Both organic farmers who do not use pesticides and conventional farmers who do use pesticides rely significantly upon commercial sources for the decisions on whether or not to spray. With organic farmers, the survey revealed that 84 percent credited an organic farming products salesman for their decision to adopt organic methods, while only one percent listed the extension service. Other organic farmers and publications received 72 percent and 61 percent respectively. In the case of conventional farmers, farmers credited their pesticide use information coming from farm magazines (94 percent), pesticide labels (90 percent), other farmers (67 percent) and chemical dealers (60 percent). County extension personnel and extension publications received ratings of 47 percent and 34 percent respectively.

The data once again confirm the need for greater attention to encouraging private sector development of the IPM industry, if a "least-is-best" pesticide strategy is to be effectively implemented by farmers.

Dr. Rosmarie von Rumker concluded the *Farmers' Pesticide Use Decisions* report by referring to the impediments which deter farmers from getting unbiased individualized crop protection information. She theorized on how the "substantial number of growers" who seek pest management advice can best obtain it.

"Pesticide industry representatives are caught in a dilemma. As documented in other parts of this report, the pesticide trade has thus far been the predominant source of pesticide information for growers. Many individuals in all echelons of the pesticide industry are very receptive to the idea of shifting emphasis from the selling of pesticides to the selling of crop protection. However, the proposition to provide services *instead* of chemicals, rather than in support of the sale of chemicals, is quite foreign to the basic policies and philosophies of many chemical companies. For most representatives of the pesticide industry who work where the 'rubber hits the road,' rewards such as raises, promotions, Christmas bonuses and/or commissions still appear to be geared to sales volumes in terms of quantities.

"Most (but not all) extension workers believe that it is beyond the scope of the federal/state cooperative extension service to furnish specific crop protection advice to individual growers on a regular basis. They point out that the extension service does not have the personnel to check more than a limited number of fields on an irregular basis and suggest that there may be a need for independent, private-enterprise crop protection consultants. The question is whether such enterprises would find sufficient economic incentives in working on insect and weed problems of midwestern field crops."

As exemplified in the next section, the new "IPM Industry," there is clearly more activity in the private sector now than when Dr. von Rumker conducted her research five years ago. However, the same barriers she notes above are still very much in existence today.

Several suggestions for motivating growers *not* to use pesticides can be drawn from the summary of Dr. von Rumker's report which is still extremely pertinent at this writing, even more so considering the additional research data on pest resistance, societal health hazards and energy constraints resulting from excessive pesticide applications.

Economic incentives are needed to increase the numbers of independent crop protection advisors. Dr. von Rumker found that many farmers interviewed were interested in receiving individualized, specific advice on crop protection, and about one-half of the respondents expressed a willingness to pay a fee for such a service. While this demand for crop protection information is gradually being met by a growing number of private IPM consultants, a tremendous gap still exists between farmers desiring unbiased assistance in crop protection, and available IPM advisory services.

Unbiased information about pesticides must be communicated to farmers. Growers need information on economic and other side effects from the unilateral use of pesticides over the long term and on interactions between various insect, weed and disease control measures. Farmers should be informed about the individual and collective effects of the chemicals on soil microflora and -fauna and on the long-term fertility of the topsoil.

As agricultural producers see how they can personally benefit from a reduction in pesticide use, they will increasingly be interested in independent crop protection advisors and unbiased information about pesticides.

The New "IPM Industry"

Chapter 4

Companies and Consultants Offering IPM Services

About ten years ago, a group of entomologists met together in the barn behind the Rincon Vitova Insectary in southern California. The purpose of the meeting was to discuss the development of a concept for an alternative approach to pesticides.

The group named itself the Association of Applied Insect Ecologists (AAIE). The AAIE has grown into an organization with several hundred members and over 20 independent businesses that provide consulting services in pest management to farmers (and in some cases, to cities and other urban and state institutions).

Basically, these early applied insect ecologists believed that the pest control advisor should be an independent person or company obtaining its income directly from the people served. This sentiment still pervades the group. Chemical company salespeople are not excluded from membership (since some new independent pest managers may come from this group of people), but the primary emphasis is on providing pest control advice that does not depend on ultimately selling a product. (This relationship of the chemical salesperson to the farmer is also being challenged through bills recently introduced into the California state legislature.)

Each pest management advisor has a strong conviction about the superiority of the kind of service he offers the farmer as opposed to that offered by the chemical salesperson. A good example is provided by Louis Ruud, who heads California Agricultural Services (Kerman, CA). Louis became involved in pest management when he was a cotton gin clerk who did substitute teaching in the off-season. He has great respect for farmers and credits them

This account of the Association of Applied Insect Ecologists, and some of its members, was written by Dr. William Olkowski and Helga Olkowski, co-directors of the Center for Integration of Applied Sciences of the John Muir Institute in California. Their programs to manage plant-pest-human interactions in urban areas are described elsewhere in this book.

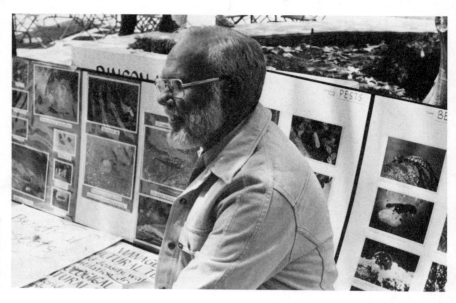

Everett Dietrick helped to pioneer commercial developments in Integrated Pest Management with his company, Rincon Vitova Insectaries.

with educating him about farm operations. Although California Agricultural Services has been primarily a pest management firm with wide experience, he sees the need for a total management service and is moving to provide this. He calls this service Integrated Farm Management. Over the years he has provided clients with advice for more than 15 different crops (cotton, melons, alfalfa hay, almonds, walnuts, peaches, ginger, apricots, sugar beets, black-eyed peas, potatoes, artichokes, seed lettuce, bell peppers, cauliflower and some minor specialty crops). Louis has many stories to tell about savings he has earned for growers. These savings resulted largely from reducing the number of treatments. The more he saves his clients, the better his reputation and the more farmers who hire the services of his company.

Larry Bowen, head of BoBiotrol (Merced, CA), also offers IPM services in southern California. Larry started in the farming community by helping organize a co-op of 200 members. He prides himself

on his empathy for the grower. His first pest management experiences were as a field checker in cotton. Over the years he has worked on many crops, and has developed his company accordingly. BoBiotrol now has had experience with fly control on dairy farms, with almonds, cotton, tomatoes, table and wine grapes, alfalfa hay and seed, walnuts, cole crops, lettuce, bell peppers, and even the new crop kiwi. Part of BoBiotrol's operations involves the rearing of the parasite *Pentalitomastix plethoricus* for release against the navel orange worm, a pest of almonds. Larry's company also raises lacewings which are sold through the mail.

Another company which combines pest management advising with rearing of natural enemies is headed by Ibrahim Michael (Fresno, CA). Michael raises the red-scale parasite *Aphytis melinus,* and also *Trichogramma* species which attack various moth and caterpillar pests. Ibrahim stresses the fact that he offers a complete service for the farmer. This includes disease management, irrigation and soil and plant tissue analysis in his own laboratory. He now provides services to citrus, grape, fig, almond, alfalfa and cotton growers.

Another AAIE member is Ron Inman, Agricultural Testing and Entomological Survey Service (Temecula, CA). Ron specializes in citrus and avocado. He stresses a holistic service orientation that also encompasses disease, soil, and plant tissue analysis, and land suitability studies for farmers. He stresses the importance of the natural enemies of the pests and has an excellent slide show to teach others how to recognize the beneficial insects.

To understand the competitive setting in which AAIE consultants operate, one needs to know that these independent business people who work solely for the farmer are still greatly outnumbered by "consultants" who work for chemical companies. Dr. Robert van den Bosch of the University of California estimates that there are about 2500 chemical company salespeople in California as compared with only 25 or so independent IPM advisor companies.

Everett Dietrick of Rincon Vitova Insectaries (Oak View, CA) has been raising insects since 1952 and has had experience with a long list of crops: cotton, alfalfa, sugar beets, sorghum, barley, citrus, avocado, tomatoes, strawberries, broccoli, cauliflower, cabbage, potatoes, truck farmed vegetables of all sorts, apples, pears and greenbeans. Estimates of total acreage being field checked by Dietrick's company vary between 10,000 and 15,000 acres. Dietrick adds that the acres directly under his advisory management are only part of the story. His insectary has had wide impact nationally and even internationally, with several hundred thousand acres being affected.

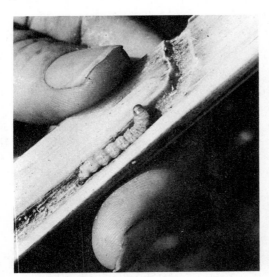 

A cross-section of an infested sugarcane stalk (left) shows a sugarcane borer larva and the enlarged exit hole it makes to accommodate the emerging moth. An adult Cuban fly (right) prepares to deposit maggots in the entrance hole of a borer; the maggots will travel up the tunnel to attack the cane borer larva. Scientists are raising the Cuban fly as a biological control weapon against sugarcane borers.
(Agricultural Research, USDA)

A discussion of how IPM consultants get started can go in a number of directions. The pattern Dietrick has seen is that many present-day independent IPM advisors first started out as employees of established consultants. In his case, the basic support for these assistants comes out of the income from the insectary business. After learning the trade, they see greener pastures and go off to start their own businesses. In the process, they take many of the original clients with them. When asked how he feels about this, Dietrick points out that he is then able to sell beneficial insects to the "new" companies. He realizes that this competition endorses his work and at the same time forces him to move into new areas where the value of IPM hasn't been established.

John Nichelson (Shafter, CA) has been checking field cotton for 28 years. He thinks that he was probably the first to offer IPM services to the farmer on a commercial basis. Today, he estimates his company must be responsible for about 70,000 acres.

John emphasizes the fact that in many cases some pesticides are needed in modern commercial agriculture. Although using the services of an IPM consultant invariably reduces dependence on pesticide use, it rarely can eliminate it. Since plant varieties, soils, watering and fertilizing practices, timing and methods of many farm operations such as pruning or cultivating can all affect pest problems, the more an IPM consultant knows about farm management generally, the better advice he can give on managing pest problems.

There is a difference between private IPM consultants and university employees who also provide pest management advice. Although both may have a desire to help the farmer, the independent consultant needs to perform well in order to survive. The final judge is the client-farmer who is impressed by the bottom line—profits.

The university employee has another boss and often a different ultimate goal—research and information. The consultants live and work in the communities they serve, while the extension agents and university researchers often live actually or at least conceptually in the intellectual environment of the institutions that support them. Much useful information and basic research are provided by the entomologists of the academic community, but it is the independent consultant who works closest to the crop systems, gets to see more aspects of the pest situations and consequently has a finger on the pulse of the farm.

—William and Helga Olkowski

Full professional membership in the AAIE requires a Bachelor's degree in natural sciences and four years full-time consulting experience. Affiliated, student and sponsor memberships are also available.

The members are consultants who practice independent pest management on a professional fee basis for individual grower clients. Typically, the consultant practices as a small group or as an individual. AAIE lists the following objectives:

1. Represent consultants engaged in professional Integrated Pest Management and gain support of those people and/or organizations striving to achieve optimum environmental quality.
2. Promote the study and recognition of Integrated Pest Management, including entomology and related fields.
3. Maximize food and fiber production with a minimum pesticide load on the environment.

4.   Provide a forum for the interchange of information and ideas to implement applied techniques and knowledge.

The address of the Association of Applied Insect Ecologists is 10202 Cowan Heights Drive, Santa Ana, California 92705.

———————————————

The Southern Alliance of Agricultural Consultants Association estimates that there are about 300 full-time, independent entomology consultants servicing field crops in the southern United States. According to the Association's president, Mills Rogers, consultants are often concerned with soil fertility and water management as well as pest control. A report in *Progressive Farmer* (April, 1978) by Del Deterling quotes one member saying that "we try to get with our customers before they even plant. We feel we can make him

as much money during the winter as in the summer by helping him plan varieties, planting dates and cultural practices that influence insect problems." A Southern Alliance spokesman estimates that the fee for cotton checked two times a week ranges from $3.00 to $6.00 an acre.

More than one-half of the Mississippi cotton acreage—about 800,000 acres—is under supervision of private consultants.

"By 1977, nearly all of the field crop consulting firms in Kansas plan to have expanded both their geographic ranges and the acreage served within their present units," reported Dr. Donald E. Mock of Kansas State University. "Some will probably double in size within the coming year. Similar expansion is occurring outside Kansas boundaries by these and other consultants. Additional consulting organizations will undoubtedly enter the competition."

Grower-Owned IPM Services

Non-profit corporations—usually some form of farmer-organized cooperatives—are becoming major deliverers of IPM services. One-half of the IPM projects in Georgia, for example, are said to be operated by non-profit, producer-controlled organizations. The USDA publication, "Establishing and Operating Grower-Owned Organizations for Integrated Pest Management," details the workings of these arrangements.

The Edgecombe Spray Program, Inc. (Tarboro, North Carolina) is one of eight spray groups organized between 1968 and 1974. During the summer, Edgecombe shares a full-time manager with the aerial applicator, and relies heavily on extension

personnel for pesticide recommendations and scout training. Through annual contracts with each farm member, Edgecombe Spray assumes full responsibility for controlling boll weevils and bollworms on cotton. To fulfill this obligation, Edgecombe Spray develops a control program, purchases the chemicals, hires and supervises the scouts, and contracts and assumes responsibility for the aerial application of pesticides and defoliants. Scouting usually saves about two chemical applications each season. Information on insect populations provided by scouts determines when insecticide applications are started.

The group practices communitywide insect control and seeks a high level of participation. Once spraying begins, it usually continues for the entire season at regular intervals of about five days, says the USDA publication.

The Growers Pest Management Corporation (Casa Grande, Arizona) operates a multicrop pest management and advisory service, with the farmer deciding on the control to use. An agreement, signed by the corporation and each member for crops entered into the program, covers fee payment and membership.

The Safford Valley (Arizona) Cotton Growers Co-op, Inc. was organized after an unusually heavy infestation of pink bollworm. The pest manager consults on pest controls, hires and supervises the scouts, supplies such materials as sex lure traps and makes recommendations for chemical applications.

Pest control contracts between the Co-op and members specify services and charges. Costs are deducted from cotton proceeds at the season's close. In this manner, Safford Valley supplies the IPM program and related services to its members at the cooperative's costs.

Servi-Tech, Inc. (Dodge City, Kansas) is a technical service organization, owned by farmers indirectly through 15 local farm supply cooperatives in western Kansas. Each technician serves about 15,-000 acres, and during the growing season has an assistant. Both have trucks, equipped with hydraulic soil probes and phones.

"Servi-Tech was incorporated as a separate entity, apart from the sponsoring cooperatives, in order to reduce the possibility of conflict with the pesticide sales department," notes the report. "This separates technical assistance and chemical sales in the farmer's decision-making."

A somewhat different category is illustrated by the Texas Pest Management Association (TPMA), an umbrella organization made up of 20 county or multi-county pest management units. TPMA will charge local projects a 15 percent administration fee, and the extension service will continue to provide training. Dr. Ray Frisbie of the University of Texas believes that TPMA "takes extension out of the business of running pest management programs and lets it concentrate on its assigned educational role."

The Visalia (California) Co-op Gin provides scouting and plant growth analysis services at no charge to its 126 growers who average about 12,000 acres of cotton. The co-op is reported to recoup scouting costs from sales of insecticides at competitive prices through its fertilizer and chemical division.

Actually, the delivery of IPM services by representatives of the pesticide industry is a critical issue. To many observers, such developments immediately create a "conflict of interest." It also places the "independent" pest management consultant at a disadvantage, since he or she earns no commission or bonus directly from his chemical company employer.

Traps and Beneficial Insects

Through their company, New England Insect Traps, Ken and Patty Spatcher of Colrain, Massachusetts sell two types of fly paper-like devices. Both traps are used in New England apple orchards for monitoring such pests as the apple maggot, tarnished plant bug and European sawfly. The Spatchers have manufactured and sold 1,200 traps during the past two years—mostly to members of the Massachusetts Fruit Growers Association.

The white rectangular trap, which attracts both the tarnished plant bug and the sawfly, is simply a cardboard rectangle painted a specific shade of white and given a coat of Tangle-Trap, a product of the Tanglefoot Company. The Spatchers coat the cardboard with the paint and Tangle-Trap, then add two lengths of wire for hanging from the trees.

A similar process is involved for the sticky red sphere traps. Wooden croquet balls, red enamel paint, Tangle-Trap and ammonium acetate are combined to trap apple maggots. A hook-and-wire piece complete the manufacturing process.

Both traps work primarily because of their color which is especially attractive to the particular insects. The light reflected by the devices corresponds to the light reflected by the plant part that the pest feeds on or mates in. With the red spheres, a smell stimulant in the form of ammonium acetate serves as a further attraction. An insect landing on either trap remains stuck in the Tangle-Trap until it is removed and counted by the grower. By averaging the number of a specific pest found on the traps, a grower can determine the approximate pest population in his orchard.

Ken Spatcher of New England Insect Traps demonstrates how traps used in apple orchards are constructed. Wooden croquet balls are covered with red enamel paint to mimic nearly red apples, dipped in Tanglefoot and hung to dry.

A cardboard rectangle is painted a specific shade of white to imitate opened apple blossoms, and then spread out to dry.

The cardboard is coated with Tanglefoot and then hung with wire on the tree to attract the tarnished plant bug and European apple sawfly.

Making Traps for Apple Orchards

(photographs by Steve Suleski)

At its insectaries in Nebraska, Gothard, Inc. rears the moth egg parasite, *Trichogramma*. The firm sells to dealers who generally offer them to their grower customers along with the service of liberating them and field checking. The *Trichogramma* are almost exclusively used on cotton, corn and tomatoes. Writes Jack Gothard, Jr.: "We are a young developing business with limited funds; therefore, growth has been limited. But as our production grows, then also will the list of beneficial insects reared grow. We will possibly get into the retail end of the biological control business on a small scale this year."

The commercial operations of Gothard, Inc. in Canutillo, Texas consist of wholesale supplying, and servicing cotton, corn and peanuts. The firm also sells *Trichogramma* for home gardeners and persons with a few fruit trees, etc. Writes Tom Gothard:

"We encourage our customers to refrain from applying pesticide as long as possible. When a customer feels that they can not stand to live with a little damage or the situation cannot be handled with releases of *Trichogramma,* we can work the releases of insects into a suitable schedule with pesticide. We have found that beneficial insects do not work as well where chemical controls are being used at the same time.

"The use of several different types of insects or other natural agents simultaneously is becoming more feasible due to the number of these becoming available on the open market. Although we propagate only one item, we have and do experiment with the use of other available insects like lacewings for controlling aphids in pecan orchards."

More People Needed to Rear and Sell Beneficial Insects

Dr. Richard L. Ridgway, an entomologist with the USDA's Agricultural Research Service in Beltsville, Maryland, is another professional who sees expansion in IPM management opportunities.

Natural biological controls in insect pest management will expand significantly if they can be more effectively managed, Ridgway maintains. "A wide range of useful beneficial insects exists in nature. However, adequate numbers are not always available in nature to provide desired control levels." Ridgway continues:

"In highly developed agriculture, dependable methods of insect control are essential. A farmer cannot effectively manage today's intensified agricultural operation in the United States unless he has highly dependable tools. For naturally occurring beneficial insects to provide the levels of pest control that modern agriculture demands, they often need to be supplemented."

One way to insure adequate numbers of beneficial insects is to mass rear and release them. This approach, often called augmentation, means adding enough beneficial insects to re-balance nature. And there are job opportunities in this area. According to Ridgway:

"There is a small, modest commercial industry and a few small cooperatives in the United States rearing and selling beneficial insects. As our knowledge continues to expand, augmentation can be expected to increase."

"Some 20 different beneficial insects for use in biological pest control are available commercially from about 25 firms," Dr. Ridgway said. "There is experimental evidence that a number of these beneficial insects can, when reared and released in large numbers, provide the desired level of pest control. Using green lacewings to control mealy bugs on pears, bollworms on cotton, and aphids on greenhouse flowers; and the egg parasite, *Trichogramma,* to control caterpillars on a number of crops are just a few examples."

The major task ahead for research is to develop less expensive methods of rearing and distributing parasites and predators, according to Ridgway. Also, he said that additional organizational entities are needed to implement the augmentation approach to insect control more effectively.

"The rate at which augmentation expands, particularly in the United States," Ridgway believes, "will depend largely on society's willingness to make financial and organizational adjustments favoring the use of biological controls."

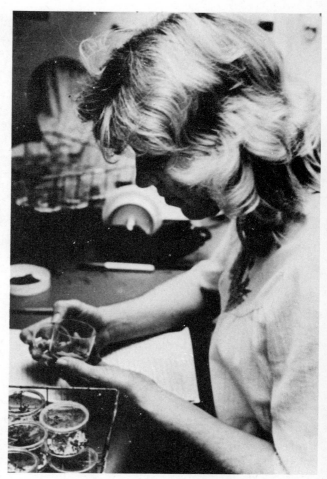

Insect rearing is essential for conducting research on biological control agents and for augmenting populations of beneficial predators. This technician is recording the amount of feeding done on different kinds of plants by the velvetbean caterpillar.
(Biological Control of Insects, Research Unit, USDA, Columbia, Missouri)

Chapter 5

Conflict of Interest
And Other IPM Barriers

"The farmer is not stupid. He will not buy and apply more pesticide than he needs for maximum profits and crop yields." Words to that effect seem to be spoken whenever the topic of a commercial "support system" for IPM comes up. The debate arises over the question of a *conflict of interest* when members of the pesticide industry—salesmen or applicators—are at the same time offering their services as advisors in Integrated Pest Management programs.

Since IPM represents a strategy for reducing pesticide use, the incentives for *independent* IPM consultants and pesticide salesmen are vastly different.

One self-employed IPM consultant recently told us how he does his best to get a farmer to understand that there is a natural enemy complex at work in his fields, and why the farmer must understand the value of habitat management. Once the farmer accepts the principle of natural balances, then he will try to see what interferences there are. "But getting a farmer

away from chemicals is something like getting an addict off drugs. It's not easy," the consultant says.

The *conflict of interest* issue is a complex one. Most often, it is associated with public policy decisions made by government officials. According to a February, 1978 column by Jack Anderson in the *Washington Post,* a startling number of officials in the regulatory agencies have financial interests in the companies they are supposed to regulate. "An official in the EPA's pesticide division owns over $10,000 worth of Exxon, Gulf and Texaco stocks. All three companies produce pesticides that are regulated by his office."

That is one form of conflict of interest.

But there is another form of conflict of interest which is far more insidious and far more destructive. This conflict relates to the key question: Who advises the farmer about how to best use Integrated Pest Management?

Despite all the rhetoric about how effective IPM is, on how many millions of acres it is being used successfully and what promise it holds for improving farmers' profits and reducing pollution hazards, many qualified observers see substantial barriers. Two persons who have written incisive analyses of the problems in expanding IPM programs onto United States farms are Darwin C. Hall and W.R.Z. Willey.

Dr. W.R.Z. Willey is a staff economist with the Environmental Defense Fund in Berkeley, California. His paper, "Barriers to the Diffusion of IPM Programs in Commercial Agriculture," clearly delineates the major issues which must be resolved if IPM is to have a major impact on pesticide usage.

At the outset of his analysis, Willey sets forth these preconditions for the application of IPM:

1. A specific crop/pest IPM technology must be available. Such technologies would include pheromones, beneficial species, detection and sampling methods, as well as narrow-spectrum pesticides.

2. IPM applications require scientific information on pest and beneficial species, alternative controls, etc. Data on a specific agro-ecosystem can only be obtained from observation in the field.

3. The technology and information must be distilled into decisions by an *objective* practitioner. "In the end," Willey stresses, "sound professional judgment is crucial to successful application. . . . Objective means that the judgment of the pest control practitioner—whether he is a grower, a farm advisor, a chemical salesman or an inde-

pendent advisor—be concerned solely with the optimum performance of the IPM program. Such optimum performance is in the grower's interest since it minimizes pest control costs while maintaining yields, i.e., the economist's notion of profit maximization."

4. IPM application must not require unbearable economic risks to the adopter; the risks involve the costs and marketability of production. IPM adoption for fruits and vegetables is seriously impeded by certain quality grades ("cosmetic" standards) which accompany marketing orders.

Willey contrasts the operating methods of information exchange offered by pesticide company personnel and independent advisors. As confirmed elsewhere in the chapter on how farmers make pesticide use decisions, most technical advice on pest control comes from field people employed by pesticide companies. The information is usually supplied free as part of the sales performance. In contrast, the independent private-sector advisor charges a fee for advice, usually with no product sale involved.

Why then, if the advice provided by independent consultants does indeed save the farmer money, isn't this fledgling industry reaching the boom period? Willey provides substantial answers:

"First, the advisors' service really involves a labor-for-capital substitution, which is exactly opposite the prevailing trend of capital intensification in agriculture and in developed economies in general. . . . Energy scarcity has not become serious enough yet to outweigh the historical advantages of capital-intensive pest control in agriculture. Capital substitution is encouraged by rising labor costs.

"Second is the fact that labor, in general, in developing economies has managed to regulate somewhat the process of capital intensification by unionizing in various ways. . . . The independent pest control advisors face the worst of both worlds—competition from a substitute (the pesticide industry) and no organized power to deal with the competition (the pesticide company, powerful and relatively organized)."

At a time when everyone from the President of the United States and the Secretary of Labor all the way down to the person-in-the-neighborhood-bar is worried about the lack of jobs, it seems that the labor-for-capital quality of IPM should be a tremendous plus. But that time has still not arrived. Willey notes that the financial resources of oil and chemical conglomerates are powerful enough to add another complex factor. "To what extent do these conglomerates," Willey asks, "through their agrichemical subsidiaries, present a serious barrier to the diffusion of independent IPM services? That is, even though the independent IPM advisors look good in terms of earnings and could in theory be considered a 'growth industry,' they earned less than one percent of total pest control revenues in California, where they

are more common than in other states. Growth of the independents' operations must occur to some extent at the expense of pesticide sales. However, the institutional rules which govern the pesticide industry were established at a time when its growth was deemed desirable by most segments of society."

(The complete text of Dr. Willey's paper will be published in *Pest Control Strategies—Understanding and Action,* edited by David Pimentel. It will be issued in 1978 by Academic Press.)

Darwin Hall, an agricultural economist at the University of California, addresses "conflict of interest" in his paper, *The Profitability of Integrated Pest Management.* His research was supported by the EPA, NSF and USDA through the University of California. Observed Dr. Hall:

"Pesticide salesmen have no incentive to monitor as closely (as independent consultants) since they would prefer to substitute pesticides for monitoring activity at the same total cost, yield and profit to the grower. Salesmen have an incentive to learn and practice IPM techniques only to the point where they can remain competitive with independent consultants. Contrarily, independent consultants can charge a per acre fee up to the savings in pesticides obtained. ... The large reduction in pesticide use which results from the efforts of independent consultants demonstrates that salesmen have a conflict of interest between profit maximization and the social goal of reduced pesticide use."

Seduction, Or Is It Rape?

Dr. Robert van den Bosch is chairman of the Division of Biological Control, Department of Entomological Sciences, University of California, Berkeley. He is a research entomologist who has helped to develop Integrated Pest Management concepts for agriculture. Several years ago, he authored an article in *Organic Gardening and Farming* magazine, entitled "The Rape of EPA." In it, he described the trouble EPA was having because "it tried to live up to its mandate" by having banned hazardous insecticides as DDT, aldrin-dieldrin, chlordane and heptachlor.

Van den Bosch contended that new laws were being drafted to erode the pesticide registration and regulation capacities of EPA. What's more, legislation proposed in 1975 would serve to certify private applicators as "competent in the use of the more hazardous kinds of pesticides."

Observed van den Bosch in his article in the January, 1976 issue of *Organic Gardening and Farming:*

"Rape is not the usual tactic of the power 'mafie' which dominate the contemporary American scene; they much prefer seduction. Seduction is clean, passive and discreet. It is what the old bureaucrat had in mind when he alluded to the compromising of federal watchdog agencies. But if seduction doesn't work or works too slowly, the violent rip-off will be used without hesitation. When EPA just wouldn't tumble to the seductive overtures of the pesticide mafia, it was made to pay a terrible price.

"... These groups and individuals have as their binding interest the preservation of the pest control status quo, which each is convinced serves its own best advantage. They don't want anyone or any agency tampering with things as they are, and they have the muscle to maul those who try.

"... It is sad to note that wide-scale implementation of integrated control will only develop out of economic chaos and human misery. But this comes as no surprise to me; chaos has been the precursor to virtually all existing programs. Man, a vain and rapacious fool, insists on doing things the hard way!"

"We Just Missed The Boat"

A University of California scientist has admitted "we just missed the boat" on the devastating effects of DBCP in the original research conducted more than 20 years ago. DBCP is the pesticide which has been blamed for causing sterility in workers and which may cause cancer as well.

"Quite frankly," Dr. Charles Hine told the *San Francisco Examiner,* "we didn't take it as seriously as we should have. In fact, nobody paid much attention to it." Hine was one of four researchers who, under contract to Shell Chemical Company, a major DBCP producer, performed animal studies back in 1952. He said the first clue to possible sterility came in 1958 when the researchers observed shrunken testicles in the laboratory animals.

Editorial writers around the country are more forcefully raising the biggest question: How many other toxic chemicals have similarly infiltrated the environment, their effects hidden, their properties insufficiently analyzed, their dangers unappreciated? "The case," writes *The New York Times,* "is a chilling reminder that our environment remains saturated with chemicals whose dangers may not yet be recognized, and our workers remain ignorant of the daily hazards. . . . Must we continue to make such belated discoveries?"

The discoveries may continue to be belated, unless activists force a closer look to be taken at the Pesticide Lobby. This problem is the subject of an article in *The Elements* (February, 1978), "How the University of California Organized the Pesticide Lobby." Here are excerpts of that article written by Becky O'Malley and Norman Wirt of the Pacific News Service:

"Officials of the University of California cooperated with pesticide manufacturers and agribusiness organizations to develop a pro-pesticide campaign aimed at 'protection of the California agriculture and food industry from all forms of illogical and nonscientific consumerist attack,' documents obtained by Pacific News Service reveal.

"A meeting called in 1970 by then-California Farm Bureau President Allan Grant mapped a strategy to defeat 'the effort to ban or seriously restrict the use of agricultural chemicals, pesticides, etc.,' as his letter of invitation stated. In attendance were representatives of big growers' associations, executives of several agricultural chemical corporations and faculty and staff from the University of California (UC) and its agricultural extension service.

"These sessions continued for four years. They eventually evolved into the 'California Educational Foundation for Agriculture and Food Production' (CEFAFP). The new information about CEFAFP, which has never previously been made public, has sparked calls for reevaluation of the historically close ties between land-grant universities like UC and big agriculture and chemical manufacturers.

" 'These new revelations show UC agricultural officials so deeply in bed with big business interests that people should start calling for a major investigation and house-cleaning,' said UC-Berkeley Physics Professor Charles Schwartz, a leading proponent of a rigorous conflict of interest code for university personnel who also work for commercial interests."

Fulfilling the Special
Interests of Society

The USDA publication, *Biological Agents for Pest Control,* includes these enlightening observations on social benefits versus private benefits from IPM use:

To society, the short run benefit of biocontrol agent use is the value of the food and fiber, and the cost is the expense of the biological method as a portion of production expenses. The advantages to society of using biocontrol techniques instead of other methods of pest control could include reduction of environmental and human health hazards. These are real and must be included in any societal accounting of production expenses.

Because the actions of an individual seldom affect the food and fiber delivery system or the state of the environment and human health, there are few incentives for individuals to attempt to include both long term considerations and societal advantages in benefit/cost calculations. However, society as a whole is affected by these considerations. Thus, the economic problem of implementing biocontrol involves overcoming difficulties of measuring costs and benefits and inducing both individual and societal actions.

In many cases this means that society must effect institutional changes such as establishing pest control districts to make it economically beneficial for producers to use biological methods. For example, in a California agroecosystem, lygus bug populations develop on but do not harm safflower; yet they do harm cotton, to which they migrate. Currently, safflower growers have no reason to control lygus bugs, and cotton growers are presumably disinclined to compensate safflower growers for the cost of taking care of the cotton growers' problem. Biological control would be an ideal solution for this type of situation and could be implemented under the auspices of a pest control district.

. . . In the context of program evaluation, key research areas related to benefit-/cost analysis can be identified. First, benefit/cost analysis can assist in identifying pest management problems that may be economically solved by the use of biological agents, either alone or in combination with other tactics. The reduction in control costs, off-target social and environmental costs, and the public costs of monitoring and enforcing must be sufficient to justify the research, regulatory and implementation costs of using biocontrol agents. Methods are needed for determining which of the available biocontrol agents are likely to have potential economic, social and environmental benefits greater than their probable costs.

Hopefully, the assessment methodology will become sufficiently sophisticated to incorporate the various unique aspects of biological control methods.

... Second, certain problems associated with benefit/cost analysis of using biological agents for pest control deserve special attention:

1 Fewer off-target social and environmental costs are associated with biological control as compared to certain chemical control methods. These advantages need to be evaluated in terms that can be incorporated into benefit/cost analysis. In some instances, economic research will produce methods for deriving adequate dollar estimates; in others, methods for determining upper and lower limits of impact will result.

2 At present, it is difficult to determine the true costs of biological control, whereas costs of chemical control are more easily determined. For example, in assessing the benefits and costs of pest scouting programs, it is difficult to determine the extent to which some participating extension personnel would have provided the same or very similar services to growers had there been no publicly supported pest scouting program. A related problem in evaluation is related to the fact that many biological agents cannot be patented, with the result that there is limited incentive for private producers to make such agents available in the needed quantity and variety.

3 Because knowledge cannot be owned and contained, the benefits of its use are difficult to quantify. For example, some growers participating in pest scouting programs use the knowledge from that program in managing fields that are not in the program. An assessment of only those fields in the program would understate the benefits; if fields in the control group included fields in which "leaked" knowledge was applied, the relative benefits of the pest scouting program would be further understated.

4 Effective pest management and other farm management practices such as fertilization, irrigation, pruning, etc., are interac-

tive. Tillage practices profoundly affect pest situations. Adequate methods are needed to evaluate the benefits and costs of major interactions. For example, it has been difficult to determine whether the larger yields of cotton by growers in California's San Joaquin Valley who utilize independent pest management consultants are attributable to the differences in pest management strategies or to differences in other management practices. The dollar value of the larger yield is several times greater than the savings in pest management costs, so the problem is not simple.

5 Growers are interested in the stability of relatively high profits. Some biological controls may be more risky due to their greater dependence on environmental conditions. On the other hand, all but the most selective chemical controls have greater risks of secondary resurgence and reemergence than do biocontrol agents. Methods for evaluating and incorporating the significance of the various types of risks to on-site decision-makers need to be developed for inclusion in benefit/cost analyses. Also needed to be incorporated into the benefit/cost analysis is the difference in risk of starting with biological approaches and generally having the option of alternative approaches subsequent to biological control failure, as compared to starting with chemical approaches and generally *not* having the option of switching to biological controls due to incompatibilities and time.

6 The time parameters for biological control and chemical control are grossly different. For example, use of a biological control agent may entail high initial costs, including crop losses for a short period, until it becomes self-sustaining; chemical control may entail lower initial costs but require repeated applications. This results in the classic economic problem of how to compare costs in different time periods, that is, determining the social rate of interest. This issue is further clouded in the case of pest management due to the mixture between private and public costs and benefits and private and public implementation roles. Appropriate methodologies for encompassing this issue can be developed for application to pest control program evaluation.

The above observations made by the contributors to the USDA report support the more specific criticisms of Dr. Hall, Willey and van den Bosch. Whether or not the points are made in emotional tones, the fact is that *substantial and varied barriers to IPM do indeed exist.* When *conflicts of interest* are layered on top of those barriers, the odds against an effective "least-is-best" pesticide strategy become overwhelming. And the incentives of the free enterprise "better mouse-trap" cannot come into play.

Chapter 6

Preparing for a Job in IPM

In 1972, the United States Office of Education awarded a contract to Kirkwood Community College in Cedar Rapids, Iowa for the development of a *Curriculum Report on Integrated Pest Management*. The report was issued, but unfortunately had little impact since it was barely disseminated. (Copies are available today via the cumbersome mechanism of ERIC, the Educational Resources Information Center in Washington, D.C.)

The teacher's role in career preparation is outlined at various educational levels. In the high schools, two types of programs are envisioned to prepare students for IPM careers. One type is referred to as "in-school" where all of the skills and related knowledge are studied in the school laboratory, set up for this purpose. The other type is called a "cooperative" program where the student acquires related knowledge in school, and skill training on released time in a real job setting.

Reference is made to pre-professional and para-professional teachers:

"Larger high schools will no doubt provide a preparatory course for post-secondary career education in Integrated Pest Management for those students whose career choice includes a decision to pursue further education after graduation. Some schools may establish a program with both pre-professional and vocational students doing their career preparation in school, on cooperative jobs, or both. Smaller schools may even combine IPM with other agriculturally related programs in a general post-secondary preparatory course.

Employability is the key to the preparation of individuals for para-professional careers in Integrated Pest Management. Whether program participants enter the world of work directly upon completion of para-professional level education and training or continue into professional level education, they must exit this stage with a high level of technical competence and a concentration of knowledge ranging from the broad-based principles of pest management to the specific applications of these principles. Much of the knowledge previously gleaned by students from courses in pest management should be given practical application here through the development of skills indigenous to specific occupations or families of occupations within the area of Integrated Pest Management. This critical need to augment knowledge with employable skills necessitates a high level of teacher involvement and individualization of programming. Para-professional teachers must, therefore, possess sophisticated skills inextricably linked to specific technician level occupations. These skills should have been developed through actual on-the-job experience as this direct experience will give the teacher an important perspective from which to impart many of these skills to students who will usually face employment, or possibly continuing education, or even both, at the culmination of para-professional education.

Essentially, the para-professional teacher's role is tied directly to these objectives:

1. The graduate should be prepared to take an entry level job as a technician and perform successfully in all occupational capacities.

2. The education and training, when combined with a reasonable amount of work experience, should facilitate the graduate's advancement to positions of increasing responsibility.

3. The foundation provided by the education and training should be broad enough to allow the graduate to pursue further educational and training opportunities at a later date."

For IPM career preparation, the Curriculum report divides jobs into three main categories—applicators, advisors and supervisors. The educational requirements range from high school to graduate education at a university.

High school and vocational training, as well as Comprehensive Education Training Act programs, can be concerned with training individuals for the following jobs: applicator, applicator assistant, field sweeper or scout, laboratory assistant, insectary technician assistant, field equipment technician and inspector trainee.

Community colleges can prepare stu-

Persons who are trained in insect identification are needed to work in Integrated Pest Management jobs. Here, Joel Grossman uses a sweep net to sample insect populations.

A research technician records oviposition to determine fecundity of a parasite wasp of the velvetbean caterpillar. *(Biological Control of Insects, Research Unit, USDA, Columbia, Missouri)*

dents for data collecting, laboratory, field and sales jobs.

Graduates in college programs can look for jobs in such areas as extension, teaching, research, advisory or consultancy, regulatory, monitoring and management.

More universities are offering pest management programs, which are, in turn, contributing support to the "IPM industry." At Simon Fraser University (in Burnaby, British Columbia), Bryan Beirne, director of the Pestology Centre, reports that enrollments have grown from six, when the program started in 1973, to more than 40 this year. "So far, all of our graduates have gotten jobs in or related to pest management: Federal Quarantine Service, Provincial Government departments in extension or control, municipal pest management (especially for mosquitoes and rats), ecological consulting, pest con-

trol, and pesticide production firms, private practice as advisory pest managers, and overseas aid agencies. A problem is that students often get such jobs before they have completed their Master of Pest Management requirements, and then delay in completing them."

Patricia Berg reports similar experiences at the University of California, Riverside, where the interdisciplinary graduate program includes an internship in the field and is aimed at "producing Pest Management Specialists qualified to work at decision-making levels."

In its introductory comments to the Pest Management Curriculum offered by the Department of Biology, Utah State University staff stress how "the demand for pest management specialists is increasing." Examples of job announcements include: pest management specialist, pest management scout, area-specialized agent —Entomology, agric-fieldman/woman, pest control consultant, agricultural management specialist, supervised control agent, extension specialist, and environment evaluation.

"The training for pest management work terminates at the BS level. It is primarily oriented toward the practical application of biological and agricultural principles, such as: identification of both pests and beneficial organisms, environmental conditions in relation to pest development, sampling, control methods, economic evaluation and decision making. Most workers with this training will be advisors for control problems associated with agricultural, public health, forest, recreational, household and urban pests."

Broad and Varied Training

"Integrated pest control and ecosystem management," Dr. Louis A. Falcon of the University of California's Entomological Sciences Department wrote, "are complex activities which usually take place in incompletely defined and inadequately understood ecosystems. A basic objective of integrated pest control is to *manage* ecosystems and *prevent* problems from developing. Where problems do arise, control actions are employed to prevent economic loss or aesthetic damage. This, however, must be accomplished while taking into consideration the ecological and sociological constraints in each ecosystem and the long-term preservation of the environment.

"To handle ecosystem management and pest control decision-making, the individuals involved must have a broad and varied training and experience built upon a firm ecological foundation. They must be capable of sifting, extracting and integrating information in a holistic way so they can make the *correct* decisions at the appropriate time. The standard education process most people are exposed to does not provide a suitable foundation for holistic thinking. Thus the impediment."

Most programs to use chemical controls for pests, whether insecticides or pheromones, are developed by industry, says Professor Richard D. O'Brien, Director of Cornell's Division of Biological Sciences. He also notes that most biological research on insect development, physiology and behavior is done by the staff of universities and units of the government. The matchup is far from balanced. "We need imaginative new ways to fuse these disciplines, and to promote exciting interactions between biologists and chemists," sums up O'Brien.

University Courses and Research Projects in IPM

A survey of entomology departments at land grant colleges throughout the United States reveals that almost all have some courses and ongoing research in Integrated Pest Management. Some, such as the University of California at Berkeley and Riverside, offer majors. Many professors who have become closely identified with IPM research and techniques use that approach in traditional entomology department courses. Some universities, such as the University of Connecticut, reported no courses in biological control. Most other universities, however, indicated an upsurge of interest from students and faculty members.

Here are some typical responses:

"We do indeed have some Integrated Pest Management programs in operation in Idaho and we have curricula at the University of Idaho at Moscow, both in the undergraduate and graduate level which deal with IPM crop protection," reported A.R. Gittins, Chairman of the Entomology Department.

At the University of Massachusetts, Professor Ron Prokopy teaches an advanced course in IPM for graduate students every other year ('79, '81, etc.) in the Fall. "We don't have as of now any courses in biological control (our faculty is few in number)," explained Prokopy. Prokopy listed the following IPM research projects at the University: integrated management of Arthropod pest of fruit, Arthropod pests of glasshouse plants, vegetable pests and biting flies.

Utah State University Entomology Professor Donald W. Davis says that most of their field-oriented entomology research is strongly oriented toward pest management. The teaching program includes a course specifically on the concepts of pest management, and one on biological control.

The Department of Entomology at the University of Maine offers one course in Economic Entomology, which deals with biological control and chemical control and explores new methods of control and their ecological implications. The Department also offers two courses in ecology: Insect Ecology, a graduate course dealing with factors affecting insect populations, and Forest Insect Ecology.

"In terms of research, nearly every faculty member and graduate student here is working on some aspect of biological control, Integrated Pest Management or insect monitoring techniques to reduce need for unnecessary pesticide sprays," writes H.Y. Forsythe, Jr., Chairman of the Entomology Department at the University of Maine. Specific bio-control research projects at Maine include work on mosquitoes, blackflies, spruce budworm and potato pests.

At North Carolina State University, one-third of the Principles of Pest Management course is devoted to elaboration of principles and philosophies, rationale and scope of IPM. A second portion is devoted to important aspects of the population dynamics of insects, phytopathogens, nematodes, vertebrates and weeds, and pertinent management strategies and tactics applicable to each pest category. The third and final portion includes in-depth presentation and study of select agroecosystems.

Oklahoma State University offers a graduate level course in Biological Control and Host Plant Resistance. It also has several research projects in greenbug resistance and alfalfa weevil resistance, as well as active biological control work against the alfalfa weevil and pecan weevil.

At Washington State University in Pullman, a full course of study in pest management should be underway in fall, 1978, according to Extension Entomology Specialist Arthur H. Retan. The course will include several classes on weed control, insect and disease control, ecological aspects of the environment and pest control. W.S.U. also has two IPM projects operating in the field: one on tree fruit pests and another on alfalfa seed. "In the fruit tree program, we are trying to improve timing of sprays where they are required and to reduce sprays where they are not needed for control of codling moth, mites, aphids and other insects involved," Retan writes. "We feel that to date we have been quite successful in both programs."

Below appears a summary of the course offerings and research projects which were reported by entomology departments.

University Course Offerings and Research Projects

University of Arizona
College of Agriculture
Professor and Head of Entomology
    Department: Dr. Theo F. Watson
Tucson, Arizona 85721

University of Arkansas
Division of Agriculture
Department Chairman of Entomology:
    Dr. F.D. Miner
Fayetteville, Arkansas 72701

Course Offerings

Undergraduate
Insect Pest Management.
Biological Control.

Research Projects

Dr. Watson is researching "Environmental Improvement Through Biological Control and Pest Management."

Course Offerings

Undergraduate
Pest Management.
Pesticides Ecology.
Insect Ecology.
Seminar (various topics).

Research Projects

Virus Diseases of Insects.
Pest Population Regulation in the Cotton Ecosystem.
Pest Population Regulation in the Soybean Ecosystem.
Virus Diseases of Pests of Forest Trees.
Host Plant Resistance in Cotton.

University of Arkansas at Pine Bluff
Associate Professor, Research Entomologist: Dr. Joseph G. Burleigh
Pine Bluff, Arkansas 71601

Course Offerings

There are no courses being taught in pest management or biological control at the present time.

Research Projects

A research project is being conducted on biological control agents of cotton. The effects of irrigation and row spacing on the population dynamics of the bollworm complex, *Heliothis zea* and *Heliothis virescens,* and the parasites of these species are the concerns of this project. Effects of pesticides on these species, are also being studied. The data from these studies will aid in the provision of some of the basic information upon which adequate pest management decisions can be made.

Auburn University
Department Chairman of Zoology-Entomology: Dr. Kirby L. Hays
Auburn, Alabama 36830

Course Offerings

A new interdepartmental curriculum in plant protection, which in reality is a pest management curriculum, will be implemented in Fall, 1978.

Research Projects

Integrated Pest Management research is being conducted on the following agricultural commodities: corn, peanuts, soybeans, cotton and pecans.

University of California, Berkeley
College of Natural Resources
Pest Management Program
Berkeley, California 94720

Course Offerings

Undergraduate

An undergraduate major in Pest Management is offered at this university which enables the student to appraise and diagnose field pest problems and to recommend economically and ecologically sound advice. Students become knowledgeable in pest population management by studying integrated control systems, instead of chemical suppression methods only, through interdisciplinary training. Following are the pest management courses required for the Pest Management major.

Introduction to the Philosophy, Ecology and Economics of Pest Management. Introduction to the systems approach to pest control, including the philosophy, goals, ecological basis, strategy and tactics of integrated control. Cropping systems, ecology, natural and artificial controls are considered.

Pest Management Methods.

Pest Management Practices.

Summer Field Course in Pest Management. Four-week summer field course in pest management principles and practices. Detection and sampling for pest and beneficial species and evaluation of damage. Experiments utilizing biological, chemical and cultural control methods.

Research Projects

The Integrated Control of Insects and Mites Attacking Strawberries: W.W. Allen.

Biological Control of the Navel Orangeworm: L.E. Caltagirone.

Integrated Control of Peach Insect Pests: L.E. Caltagirone.

The Principles, Strategies and Tactics of Pest Population Regulation and Control in Major Crop Ecosystems: Multidisciplinary project involving 18 universities; management centered at U.C. Berkeley: Carl B. Huffaker.

Biological Control of Weeds (Puncture Vine, Poison Oak, Poison Ivy): Carl B. Huffaker.

Biological Control of Field Bindweed, Yellowstar Thistle and Russian Knapweed: Carl B. Huffaker.

Integrated Pest Management for Olives: Carl B. Huffaker

Predictive Modelling of Mortality Caused by Pathogens Released into Field Populations of Insects: D. E. Pinnock.

Microbial Control of Mosquito

Portable computer terminal permits fast field access to central computer-based models for prediction and recommendation of Integrated Pest Management techniques. *(California Agriculture)*

Larvae: D.E. Pinnock.

Etiology, Epizootiology and Management of Chalkbrood Disease of *Megachile Rotundata* in Alfalfa Seed Producing Areas: D.E. Pinnock.

Characterization of Viruses Associated with *Baculovirus* formulations: D.E. Pinnock.

Honey Bee Pathology: D.E. Pinnock.

Use of Soil Factors and Soil Crop Interaction to Suppress Diseases Caused by Soil-Borne Plant Pathogens: A.R. Weinhold.

Integrated Pest Management on Alfalfa: A.R. Weinhold.

Biological Control of Aphids: Robert van den Bosch.

Biological Control of Alfalfa Weevils: Robert van den Bosch.

Biological Control of Cotton Pests: Robert van den Bosch.

Biological Control of Pests of Urban Plants: Robert van den Bosch.

University of California, Riverside
Department of Entomology
Chairman of Pest Management Program:
  Dr. W.H. Ewart
Riverside, California 92502

Course Offerings

Graduate

A Master of Science degree in Pest Management is offered here. Following are the graduate courses in the Pest Management core program.

Principles of Pest Management. Basic concepts for control of insects, plant pathogens, nematodes and weeds in various types of plant communities; their implementation in integrated systems of pest management; and the environmental, ecological and legal considerations governing their use.

Methodology of Pesticide Application. Principles, techniques and equipment utilized in the application of pesticides in agricultural fields and urban situations; includes consideration of climatic factors influencing application and environmental contamination.

Properties and Utilization of Pesticides. Properties of pesticides and their utilization in pest management systems; formulations and application methods; target-organism selectivity; factors affecting pest control efficiency; residues and their fate in the environment; and hazards and regulations.

Formulations and Implementation of Pest Management Strategies. Theory and basis for the formulation, development and integration of various types of control measures against insects, plant pathogens, nematodes and weeds; includes biological control, growth hormones, sex attractants, male sterility, resistant varieties, exclusion, eradication and chemical and physical control.

Directed Studies (one summer of work experience).

Research Projects

Environmental Improvement Through Biological Control and Pest Management: Drs. E.R. Oatman, J.A. McMurtry and I.M. Newell.

Development of Integrated

Strategies for Management of Mosquito Populations: Drs. B.A. Federici and E.F. Legner.

The Development of Invertebrate Pathogens for Use in Pest Management Programs: Dr. B.A. Federici.

Biology, Ecology & Management of Arthropods in Southern California Almond, Walnut & Pome Fruit Orchards: Dr. M.M. Barnes.

Principles, Strategies and Tactics of Pest Population Regulation and Control in Citrus Ecosystems in California: Drs. L.A. Riehl, T.W. Fisher, J.B. Bailey, P. DeBach, W.H. Ewart, J.A. McMurtry, D.S. Moreno and R.F. Luck.

Investigation of the Integration of Chemical and Biological Control of Arthropods Attacking Field and Forage Crops: Dr. V.M. Stern.

Codling Moth Population Management in the Orchard Ecosystem: Dr. M.M. Barnes.

An Integrated Approach to Pest Management in Stored-Products to Prevent Insect Infestation: Dr. R.G. Strong.

Development of Integrated Strategies for Management of Mosquito Populations: Dr. M.S. Mulla.

Urban Pest Management—Pests of the Urban Environment, Their Biology, Distribution and Control: Drs. R.E. Wagner and M.K. Rust.

Predictive Decision-Making Models in an Integrated Pest Management Program on Field Crops: Dr. V.M. Sevacherian.

Development of Integrated Strategies for Management of Mosquito Populations: Dr. G.P. Georghiou.

Control Insects & Mites of Vegetable Crops: Dr. J.A. Wyman.

Nematode Population Management in Perennial Crops: Drs. H. Ferris, I.J. Thomason and S.D. Van Gundy.

Research, Development and Use of Nematode Pest Management Systems: Drs. S.D. Van Gundy and H. Ferris.

Development of Integrated Strategies for Management of Mosquito Populations: Drs. E.G. Platzer and S.D. Van Gundy.

Control Insects & Mites of Field & Forage Crops: Dr. H.T. Reynolds.

Control of Insects and Related Pests of Floricultural Crops, Nursery Stock and Turfgrasses: Dr. R.D. Oetting.

Clemson University
Department Chairman of Entomology and Economic Zoology: Dr. S.B. Hays
Clemson, South Carolina 29631

Course Offerings

Undergraduate and Graduate

Insect Pathology. The study of insect diseases including those caused by viruses, rickettsiae, bacteria fungi, protozoa and nematodes; the effects of diseases on insect populations and the use of pathogens in insect control.

Graduate

Insect Ecology. Principles of insect ecology, population dynamics and natural regulating mechanisms of insect populations; effect of environment on distribution and abundance of insects.

Insect Pest Management. Application of ecological principles to the management or control of insect populations; major factors influencing insect population fluctuations; integrated systems including biological, cultural, physical and chemical and other techniques forming a unified multi-phasic approach based on applied ecology.

Research Projects

Insect Pest Management.

Alfalfa Insect Pest Management.

Development of Pathogens for Use in a Pest Management System for Soybean Insects.

Bionomics and Control in Cotton Insects.

Biology, Ecology and Management of Peach Insects.

Insects as Hosts and Vectors of Viruses.

Pathological Relationships between Insects and Biological Control Agents.

Control Tactics and Management Systems for Arthropod Pests of Soybeans.

An Integrated System for Suppression of the Boll Weevil.

Soybean Production and Management Simulation Models.

Development of a Grower Treatment Algorithm for Insect Pests of Cotton.

Impact of Pesticides on Fungal Pathogens in the Soybean Ecosystem.

Development and Evaluation of Soybean Cultivars Resistant to Insect Pests.

Colorado State University

Professor in Department of Zoology and Entomology: Dr. W. Don Fronk
Fort Collins, Colorado 80523

Course Offerings

Undergraduate

Principles of Insect Control. Focus is on the subject of Integrated Pest Management.

Biological Control of Insects. This course focuses only on the biological control aspect.

Insect Pests of Crops and Livestock, Insect Pests of Plants and Forest Entomology. Integrated control is discussed in these courses.

Research Projects

Alfalfa weevil control (biological control).

Control of sugar beet insects (integrated control).

Range caterpillar control (integrated control).

Pine insects (biological control).

Cornell University

Department of Entomology
Ithaca, New York 14850

Course Offerings

Undergraduate

Insect Pest Management. Introduction to principles and techniques of insect pest management as these relate to the diverse problems in contemporary economic entomology.

Biological Control. Theory and method of biological control of arthropod pests and weeds.

Research Projects

Ecology and Economics of Insect Population Control: Dr. D. Pimentel.

Studies on the Stability of Insect-Plant Associations: Dr. R. B. Root.

Biological Control of Insects—Ecology and Behavior of Beneficial and Pest Species: Dr. M. J. Tauber.

A field-sweeping device is used to identify the insects inhabiting this field.
(*California Agriculture*)

Control of Insect and Mite Pests of Woody Ornamental Plants, Shade Trees and Christmas Trees: Dr. W. T. Johnson.

Plant Resistance to Arthropods in Pest Management: Dr. W. M. Tingey.

Biology and Control of Medically Important Arthropods: Dr. E. W. Cupp.

Biology and Control of Arthropods Affecting Livestock and Poultry: Dr. E. T. Schmidtmann.

Black Fly Damage Thresholds, Biology and Control: Dr. E. W. Cupp.

Alfalfa Weevil Control and Environmental Conditions, Cutting, Management, and Insecticide Applications: Dr. R. G. Helgesen.

Vegetable and Forage Crop Insects-Ecology and Control Insect Dispersal/Migration: Dr. A. A. Muka.

Forage Insect Investigations: Dr. R. G. Helgesen.

Integrated Control of Arthropod Pests Affecting Florist Crops: Dr. R. G. Helgesen.

Pest Population Regulation and Control in the Alfalfa Ecosystem: Dr. R. G. Helgesen.

University of Delaware
Department Head of Entomology and Applied Ecology: Dr. Dale F. Bray
Newark, Delaware 19711

Course Offerings

Biological Control of Pests.
Elements of Entomology. Pest man-

agement and biological control is discussed in this course.

Economic Entomology. Pest management dealt with extensively in this course.

Research Projects

A major project, which has been going on for several years, concerns a newly discovered parasite of European corn borer eggs. Live material has been sent to the Soviet Union for testing there. Research is being conducted on alternate hosts for parasites of the Gypsy moth, and an extensive test on the control of the Mexican bean beetle through use of a parasite is underway.

The goal of the entire mosquito control research effort is directed toward the best Integrated Pest Management program possible. Dr. Bray, Chairman of the Entomology Department, describes this research, "In it we use both high-level and low-level impoundments. We also test special ditching techniques and have developed an excellent micro-marsh habitat test method for screening potential mosquito larvicides against non-target marsh species such as fish, crabs, grass shrimp, other insects, etc. These micro-marshes are located on our college farm so that we do not expose any of the natural marshes to the candidate pesticide until it is tested in the micro-marsh first. We also have a project on selecting a cold-hardy strain of the mosquito eating fish of the genus *Gambusia.*"

The University of Georgia
Department of Entomology
Assistant Professor: Dr. John All
Athens, Georgia 30602

Course Offerings

Undergraduate

Agricultural Entomology. This course is an introduction to entomology with emphasis on agricultural pests. Biological control and other suppression methods are discussed in general and as they are applied to specific pests. Concept of IPM is discussed.

Undergraduate and Graduate

Applied Systems of Insect Control. All types of insect control are considered except chemical control. Emphasis is on biological control, host plant resistance, and cultural control methodology with discussion of practical integration of these techniques (with and without chemical control) into an economical IPM system.

Integrated Plant Pest Management. IPM systems for the entire pest spectrum infesting plants (mainly insects, diseases and weeds) are studied. Pest suppression techniques for the various types of pests are integrated into a logical system for practical application by growers in this course.

Research Projects

Professor All's research, which strongly emphasizes IPM, is devoted to the study of the biopotential of arthropods in various multiple cropping systems (two or more crops on the same land) involving corn and sorghum. He has been specifically studying the ecology of conservation tillage cropping and the influence of these systems on pests and the abiotic and biotic mortality factors that affect these insects. The identification of cropping systems that have greater vulnerability to pests as well as systems that have direct control potential (cultural control) is the purpose of this research. Additional results of this work are to identify conditions that enhance or detract from biological control.

University of Hawaii at Manoa
Acting Chairman of Entomology Department: Dr. Frank Haramoto
Honolulu, Hawaii 96822

Course Offerings

Undergraduate
Biological Control of Pests.
Insect Pathology.

Research Projects

Population Ecology and Pest Management.
Infectious Disease of Insects in Hawaii.

This water-pan trap is used to snare the green peach aphid.
(California Agriculture)

Biology and Control of Hawaiian Mites.

Biology and Control of Insect Pest of Native and Cultivated Plants.

Biological and Integrated Control of Arthropod and Weed Pests.

Environmental Improvement Through Biological Control and Pest Management.

Biology and Control of Forest Tree Borers in Hawaii.

University of Idaho
College of Agriculture
Professor and Chairman of Entomology Department: Dr. A.R. Gittins
Moscow, Idaho 83843

Course Offerings

Idaho University has a curricula on the undergraduate and graduate level which deals with Integrated Pest Management (crop protection).

Research Projects

Following are Integrated Pest Management action programs at the University of Idaho:

Green Peach Aphid, Leaf roll potato program: Mr. Larry E. Sandvol, Extension Entomologist, Research & Extension Center, Aberdeen, Idaho.

Alfalfa Seed Pest Management Program: Dr. Ron Bitner, P.O. Box 1058, Caldwell, Idaho.

Peas and Lentils: Dr. L.E. O'Keffe, Department of Entomology, University of Idaho, Moscow, Idaho.

Sweet Corn : Professor D.R. Scott, Department of Entomology, University of Idaho, Moscow, Idaho.

Iowa State University of Science and Technology
Department Chairman of Entomology: Dr. Paul A. Dahm
Ames, Iowa 50011

Course Offerings

Undergraduate

Following are the core courses of the Pest Management and Crop Protection curriculum:

Applied Entomology.
Environmental Biology.
Agri-Meteorology.
Plant Pathology.
Weed Identification and Control.
Agricultural Law.
Soil Genesis and Survey.
Plant Breeding.
Special Problems in Pest Management.
Economics of Natural Resources.

Research Projects

Host Plant Resistance Studies: European Corn Borer on Corn and

Sorghum (includes artificial infestation and plant damage evaluation techniques, and genetic and plant breeding methods for developing corn and sorghum genotypes resistant to the European corn borer): Dr. W. D. Guthrie.

Orchard Pest Management: Dr. E. R. Hart.

Biology and Management of the Mimosa Webworm on Urban Plantings of Honey Locust: Dr. Hart.

Biological Control and Insect Pathology of Corn Insects (use of microsporida, bacteria, fungi and viruses for insect population suppression): Dr. L. C. Lewis.

Natural Enemy Studies (to include investigations on the interrelationships of parasitoids, predators and pathogenic microorganisms attacking pests of corn, soybeans and forage crops): Dr. J. W. Mertins.

State Project—Management of Insect Pests of Corn. Includes developing new control alternatives and improving the old to increase the efficiency, economy and environmental compatability of insect control alternatives available to the corn grower: Dr. J. J. Tollefson.

EPA-Supported Project—Development of Pest Management Strategies for Soil Insects on Corn. A cooperative project with five other Corn Belt states that will, over the next four years, put together the best possible pest management program for the three leading soil insect pests of corn: the corn rootworm, black cutworm, and wireworm: Dr. Tollefson.

University of Kentucky
Department Chairman of Entomology: Dr. B.C. Pass
Lexington, Kentucky 40506

Course Offerings

Undergraduate
Insect Pest Management.
Biological Control of Insects.

Research Projects

Integration of Biological and Chemical Control of the Alfalfa Weevil, *Hypera postica.*

Control Tactics and Management Systems for Arthropod Pests of Soybeans (Insect Pest Management on Kentucky Soybeans).

Louisiana State University
Department Head of Entomology: Dr. Jerry B. Graves
Baton Rouge, Louisiana 70803

Course Offerings

Undergraduate
Biological Control.
Fundamentals of Applied Entomology. This course specifically deals with Integrated Pest Management.

There are two curricula in Entomology, which both stress the principles of IPM. The Science curriculum is designed for students who know early in their career that they will do graduate work in Entomology. The Plant and Animal Protection curriculum is designed to graduate a student at the B.S. level who can practice IPM at the user level. Louisiana State University has used IPM principles and strategies in the control of insects in their state for the past 23 years.

Research Projects

IPM research, which has received national and international recognition, has been conducted on soybeans by Dr. L.D. Newsome and on sugarcane, by Dr. S.D. Hensley.

University of Maine at Orono
Department Chairman of Entomology:
Dr. H. Y. Forsythe, Jr.
Orono, Maine 04473

Course Offerings

Undergraduate

Economic Entomology. Deals with biological control, chemical control and explores new methods of control and their ecological implications.

Forest Insect Ecology. Concerned with the insects of the forest ecosystem.

Graduate

Insect Ecology. Factors affecting insect populations.

Research Projects

Almost every faculty member and graduate student at the University of Maine is working on some aspect of biological control, Integrated Pest Management or insect monitoring techniques to decrease the need for unnecessary pesticide sprays. Present bio-control research projects include bio-control of mosquitoes, predators of blackflies, predators of potato pests, spruce budworm pathogens *(Bacillus thuringiensis)*, Entomopthora fungi, and the budworm parasite—*Brachymeria intermedia.* This Department of Entomology reflects the State of Maine's interest concerning pesticide impact on the environment by researching these alternate methods.

University of Massachusetts
Chairman of Entomology Department:
Dr. Ron Prokopy
Amherst, Massachusetts 01002

Course Offerings

There is an advanced course for entomology graduate students that is taught every other year in the Fall (Fall '77, '79, '81) on Integrated Pest Management.

Research Projects

Integrated Management of Arthropod Pests of Fruit: Dr. Prokopy, University of Massachusetts.
Integrated Management of Arthropod Pests of Glasshouse Plants: Dr. A. Gentile of Suburban Experimental Station at Waltham, Massachusetts.

Integrated Management of Biting Flies: Dr. Edman, University of Massachusetts.

Integrated Management of Vegetable Pests: Dr. Kring, University of Massachusetts.

Michigan State University
Ecosystems Design and Management Program: Dean L. Haynes
East Lansing, Michigan 48823

Course Offerings
Undergraduate

The Ecosystems Design and Management Program includes a number of courses which cover Integrated Pest Management, soil, stream and lake ecosystems.

Research Projects

On-line Pest Management of Field and Vegetable Crop Pests.

Environmental Implications of Pesticide Usage.

Integrated Control of Tree Fruit Pests.

Laboratory and Field Studies of Pheromones of Michigan Insect Pests.

Pathology and Biological Control of Insect Vectors of Animal and Plant Diseases.

Biological Control of Selected Major Insect Pests in Michigan.

Environmental Monitoring for Pest Management Systems.

Ecology of Nematode Host-Parasite Relationships.

University of Minnesota
Department Chairman of Entomology, Fisheries and Wildlife: Dr. Milton W. Weller
St. Paul, Minnesota 55108

Course Offerings
Undergraduate or Graduate
Forest Entomology. Essence of biological control in forest systems.
Integrated Pest Management.
Current Topics. Integrated Pest Management.
Principles of Economic Entomology. Essentials of IPM as well as traditional approaches.
Experimental Ecology. Provides background essential to IPM.

Graduate Only
Insect Ecology. Provides background on populations essential to IPM.

Research Projects
(Funded mainly by the Minnesota Agricultural Experiment Station)

Host-Plant Resistance and the Integrated Control of Vegetable-Crop Insects: Dr. E.B. Radcliffe.

Host-Plant Resistance and the Integrated Control of Forage-Crop Insects: Dr. E.B. Radcliffe.

Genetics and Ecology of the European Corn Borer: Dr. H.C. Chiang.

Ecology of Corn Rootworms: Dr. H.C. Chiang.

University of Missouri-Columbia
Department of Entomology
Coordinator of Pest Management Program: Dr. Reed L. Kirkland
Columbia, Missouri 65201

Course Offerings

The University of Missouri offers a Pest Management teaching program, headed by a faculty representing four departments in the College of Agriculture. Required courses for all Pest Management majors are the following:

Undergraduate
Agronomy: Weed Control.
Plant Pathology: Introduction to Plant Pathology.
Entomology: Bionomics of Insect Pests.
Pest Management: Principles of Pest Management.
Entomology: Pesticide Chemicals.
Pest Management 198: Pesticide Application Technology.
Pest Management: Advances in Pest Management.
Pest Management: Internship in Pest Management.
Computer Science.

Biological Control of Insects (Entomology) is not a required course but strongly recommended for and taken by 95 percent of the students majoring in Pest Management.

Research Projects

Control of Medical and Veterinary Pests: Dr. Robert Hall.

Control of Forage Pests: Dr. James Huggans.

Bionomics and Control of Forage and Ornamental Pests: Dr. William Kearby.

Control of Soil Arthropod Pests of Corn: Dr. Armon Keaster.

Toxicology: Dr. Charles Knowles.

Development of an On-Line Pest Management Program: Dr. Rodney Ward.

Biological Control of Greenbug: Dr. Reed Kirkland. (Dr. Kirkland is presently rearing a new wasp species introduced into the United States from Pakistan which is proving to be effective in controlling greenbug in the laboratory.)

Montana State University
Professor of Entomology: Dr. Norman L. Anderson
Bozeman, Montana 59715

Course Offerings

Undergraduate
Concepts of Pest Management.

Graduate
Biological Control of Insects.

Research Projects

Montana State University is involved in an Agricultural Research Project concerning biological control of weeds, which is contributing to a larger Regional Research Project, W-84: "Environmental Improvement Through Biological Control and Pest Management." This research is focused on insect, mite and weed pests, and is being conducted by 11 Western Experimental Stations and the USDA.

Integrated Pest Management.

Undergraduate/Graduate
Insect Control by Host Plant Resistance.
Current Problems in Economic Entomology.

The University of Nebraska-Lincoln
Department Chairman of Entomology:
    Dr. E.A. Dickason
Lincoln, Nebraska 68583

Course Offerings

Undergraduate
Field Crops Insects.

Research Projects

Several of the departmental research projects have Integrated Pest Management as one, or more, of the research objectives. Projects relate to research on insects affecting corn, sorghum, alfalfa, turf, miscellaneous crops and livestock.

After infestations with spotted alfalfa aphids, seeded flats are checked by researchers for plant growth; the absence of plants indicates which varieties of alfalfa were killed by aphids.
(Agricultural Research, USDA)

University of Nevada
Extension Entomologist: Dr. Robert Lauderdale
Reno, Nevada 89507

Course Offerings

There is one course offered in Integrated Pest Management that may be taken by either graduate or undergraduate students. The major thrust of this course is to train students so that they may be employable in alfalfa seed pest management. The extension service of the University of Nevada is in the third year of an IPM program for alfalfa seed production.

Research Projects

Although there are no specific research projects in IPM at this university, a certain amount of investigative work is done under the alfalfa seed project.

New Mexico State University
Department of Botany and Entomology
Professor: Dr. John C. Owens
Las Cruces, New Mexico 88003

Course Offerings

A Bachelors Degree in Pest Management and the Bachelors in Agricultural Biology is offered by this department, as well as an interdisciplinary Master of Science in Pest Management. *Introduction to Pest Management* is the first academic course taken by students, and subsequent courses deal with specific areas of pest management.

Research Projects

Following is a listing of research projects on Integrated Pest Management:

Corn Grain Sorghum and Wheat Insect Pests.
Cotton Insects.
Cotton Diseases.
Livestock Pests.
Forest Insect Pests.
Rangeland Insect Pests.
Pecan Insects.
Alfalfa Insects.
Vegetable Diseases.

North Carolina State University at Raleigh
Department of Plant Pathology
Assistant Professor of Plant Pathology and Coordinator of IPM Teaching Program: Dr. Blanche C. Haning
Raleigh, North Carolina 27607

Course Offerings

Undergraduate

The Integrated Pest Management Teaching Program offers basic and advanced coursework in each of the three pest disciplines, Weed Science, Plant Pathology and Entomology, in addition to coursework in crop science, soil science, ecology and plant physiology. In the senior year, students take Principles of Pest Management and a Pest Management Seminar.

In Principles of Pest Management, one-third of the course is devoted to elaboration of principles and philosophies, rationale and scope of IPM. The science of the pesticide problem is presented along with an introduction to systems science and the roles of surveillance, monitoring, sampling, modeling. Economic and sociologic aspects of IPM are discussed. A second portion of the course is devoted to important aspects of the population dynamics of insects, phytopathogens, nematodes, vertebrates and weeds, and pertinent management strategies and tactics applicable to each pest category. The third and final portion of the course includes in-depth presentation and study of select agroecosystems: cotton, small grains, tobacco, corn-/soybeans, forest deciduous fruit crops and urban areas. Participation in the interdepartmental seminar follows this course.

At this point in time, the NCSU IPM Academic Program, which is an interdepartmental program among the Departments of Plant Pathology, Crop Science, Horticultural Science and Entomology, is still evolving. It is administered by an Advisory Committee under the chairmanship of Dr. Blanche C. Haning.

Research Projects

Ecology and Management of *Heliothis* Populations on Cotton, Corn, Soybeans and Other Host Plants.

Control Tactics and Management Systems for Arthropod Pests of Soybeans.

An Integrated System for the Suppression of Boll Weevil.

Insect Pest Management in Coastal and Esturarine Areas.

Biology and Control of Soybean and Corn Insects in Tidewater, N.C.

Identification and Utilization of Insect Resistance in Vegetable Crops.

Integrated Pest & Orchard Management Systems for Apple in N.C.

The Ohio State University
Department Chairman of Entomology:
    Dr. D. Lyle Goleman
Columbus, Ohio 43210

Course Offerings

Undergraduate
Economic Entomology.
Pesticides, the Environment and Society.
Pesticide Regulations.
Insect Pathology.
Biological Control.

Graduate
Advanced Economic Entomology.

Research Projects

Action Programs Involving Integrated Pest Management of Field Crops: Dr. Blair.

Integrated Control of Apple Pests: Dr. Holdsworth and Dr. Horn.

Management and Control of Insects Attacking Corn: Dr. Clement.

Control of Insects on Turf and Ornamentals: Dr. Niemczyk and Dr. Nielsen.

Pathogens of Insects: Drs. Hink, Stairs and Briggs.

Oklahoma State University
Department Chairman of Entomology: Dr. Don C. Peters
Stillwater, Oklahoma 74074

Course Offerings

Undergraduate
Principles of Economic Entomology.

Graduate
Biological Control and Host Plant Resistance.

Research Projects

Some of the basic material concerning greenbug resistance grain sorghum has been provided by Oklahoma State University, and in cooperation with the Agronomist there, they are hoping to have greenbug resistance in alfalfa and resistance to several pests of cotton. Active biological control work against the alfalfa weevil and the pecan weevil is being conducted, and they are working with a series of imported pests of pests of greenbugs.

Oregon State University
Department of Entomology
Associate Professor of Entomology: Dr. Ralph E. Berry
Corvallis, Oregon 97331

Course Offerings

Undergraduate
Applied Entomology. Recognition, biology and management of injurious and beneficial insects; insects and human welfare.

Seminar in Pest Management. Current topics related to IPM are discussed by students and staff interested in the broad field of pest management.

The Undergraduate Curriculum in Pest Management for Plant Protection was established in 1974 with the goal of training students in pest management. The program is interdisciplinary and requires that students take courses in weed science, plant pathology, agriculture and entomology. It is designed to provide a broad background in principles and skills involved in managing pest control programs.

Graduate
Plant Protection Entomology. The chemical, cultural and biological control of insects pests of crops; an integrated approach.

Field Methods in Insect Pest Management. Field and laboratory oriented course designed to introduce students to research methods employed in insect pest management.

Biological Control. Use of biotic agents in control and population regulation of insect pests and weeds; case history examples of biocontrol.
Advances in Pest Management. Discussion of current topics and recent advancements in insect pest management.

There is a graduate program in Entomology where a student may specialize in Pest Management.

Research Projects

Integrated Pest Management of Insects on Tree Fruits and Nuts: Dr. Ali Niazee.
Biology and Management of Soil Arthropods: Dr. Berry.
Biology and Management of Insects on Mint: Dr. Berry.
Integrated Pest Management of Alfafla Seed Pests: Dr. Fisher.
Epidemiology and Management of Insect-Borne Plant Pathogens: Dr. Clarke.
Management of Tansy Ragwort: Dr. McEvoy.
Proposed: Integrated Pest Management of Insects in Mint: Drs. Fisher and Berry.

The Pennsylvania State University
Department Head of Entomology: Dr. B.F. Coon
University Park, Pennsylvania 16802

Course Offerings

Graduate
Biological Control.

Insect Pathology.
Pest Management.

Research Projects

Integrated Pest Management Systems and Techniques for Apple Orchards.
Management of Insect and Mite Pests of Grapes.
Bionomics, Ecology and Management of Alfalfa Insect Pests.
Apple Pest Management Application Project.
Utilization of Entomophagous Insects in Pest Management Systems in Pennsylvania.

Purdue University
Department Chairman of Entomology: Dr. E. Ortman
West Lafayette, Indiana 47907

Course Offerings

Undergraduate
Biological Control. A course for non-entomology majors.

Research Projects

Pest Management in Forage Crops and Soybeans: Dr. Richard Edwards.
Corn Pest Management: Dr. Tom Turpin.
Host-Plant Resistance: Professor Richard Shade.
Population Dynamics and Quantitative Biology: Dr. David Vail.
Forage Insects: Professor Wilson.

University of Rhode Island
College Resource Development
Extension Plant Protection Specialist: Dr.
 David B. Wallace
Kingston, Rhode Island 02881

Course Offerings

Undergraduate

Introduction to Plant Protection.
Applied Insect Ecology.

Research Projects

The University of Rhode Island is presently involved in research and an extension pest management program on potatoes. These projects, under the supervision of Dr. Richard Casagrande, are concerned with the assesment, forecasting and reduction of insect and disease problems on potatoes. Research is being conducted on disease resistant vegetable varieties by David B. Wallace, the Extension Plant Pathologist, who is also establishing a Pest and Disease Forecasting Program for the state of Rhode Island.

Simon Fraser University
Master of Pest Management (M.P.M.)
Director of Pestology Center: Dr. Bryan
 Beirne
Burnaby, British Columbia V5A 1S6

Course Offerings

Graduate

The Master of Pest Management (M.P.M.) graduate program consists of courses and field work which permit the student to emphasize a particular class of pest, environment or management procedure. The M.P.M. circular (January, 1977) states that the M.P.M. degree requires completion of a minimum of six courses selected from the following in discussion with Faculty Advisors of the Pestology Center.

Biological Controls. Principles, theory and practice of the use of living organisms in the natural regulation and applied control of pest organisms.

Physical Controls. Theory and operation of physical agents and processes in natural and applied controls.

Pest Management in Practice. Special problems and procedures of pest management in different environments and of pest management agencies, interactions and communications.

Population Processes. Operation of population dynamics and behavioral ecology in determining the fluctuations and regulation of animal, especially insect, populations.

Insecticide Chemistry and Toxicology. The chemistry, toxicology, metabolism and fate in the environment of insecticides.

Biology and Management of Forest Insects. The basis is critical analyses of information on the bionomics, economic impact, ecology, etc., of representation species.

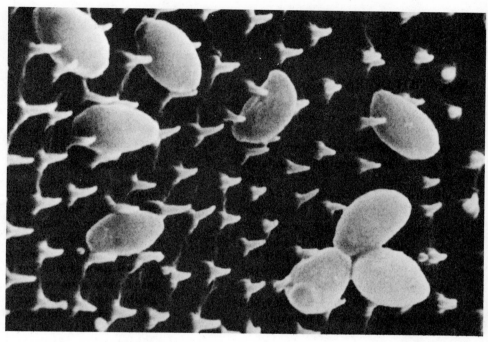

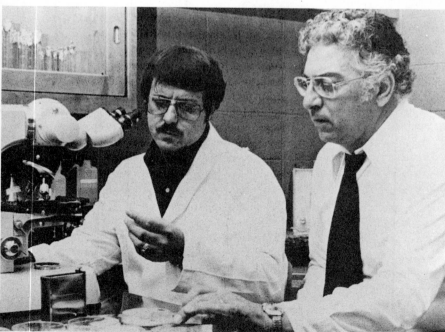

Research on the use of safe, selective microorganisms for insect control is being conducted at the USDA Center in Missouri. These scientists are studying the isolation, identification, specificity and propagation of pathogenic microorganisms as microbial insecticides. Microbial insecticides are made from fungi such as this one (condia of Namuraea rileyi), as well as from bacteria, virus and protozoa. *(Biological Control of Insects, Research Unit, USDA, Columbia, Missouri)*

Analysis of Insect Pest Situations. Critical analysis of research related to economic thresholds and control and management procedures for representative pest species, mainly agricultural.

Plant Disease Development and Control. Development of disease of cultivated crops in relation to various control and management practices.

Nematology. Host-parasite relationships of plant- and insect- parasitic nematodes and their relevance to human society.

Weed Biology and Control. Nature, characteristics, ecology, economics, control and management of unwanted plants.

Vertebrate Pests. Evaluation of the biology and social impact and the applied control techniques of vertebrates that are in conflict with human activities.

Medical and Veterinary Entomology. Analysis of problems in the management of insects and related organisms that on a world basis are important in harming directly, or in carrying diseases of, man or livestock.

Plant Disease Vectors. Biology, regulation and control of arthropod vectors of plant disease, especially aphids, and including specificity of vector-plant interactions.

South Dakota State University
Department Chairman of Entomology-
    Zoology: Dr. R.J. Walstrom
Brookings, South Dakota 57006

Course Offerings

Undergraduate

Biological control and Integrated Pest Management are taught as part of the following courses:

Introduction to Entomology.
Household Pest Control.
Crop and Livestock Insects.
Horticultural Insects.
Medical Entomology.
Insect Control Methods.

A new curriculum in Pest Management will be offered, beginning in the fall of 1978 which will provide the student with a Bachelor of Science degree in Pest Management.

Research Projects

The following research projects in this Department concern, in part, biological control and/or Integrated Pest Management:

Corn Rootworm in South Dakota.
Fly Control on Livestock in South Dakota.
Insects Injurious to Range Grasses in South Dakota.
Aspergillus Fungi Control in Birds in South Dakota.
Alfalfa Seed Insect Management in South Dakota.

The University of Tennessee
Agricultural Biology Department
Assistant Professor: Dr. Paris L. Lambdin
Knoxville, Tennessee 37901

Course Offerings

Graduate
Concepts of Insect Pest Management.

Research Projects

Insect Pest Management research is being concentrated at the present time on tomatoes, soybeans and cabbage: selective use of chemical compounds for insect pest control and their effect on nontarget organisms; effect of parasites and predators on the major insect pests; and feasibility of utilizing trap crops for control of major pests.

Texas A & M University
Professor and Chairman of Entomology Department: Dr. P. L. Adkisson
College Station, Texas 77843

Course Offerings

Undergraduate
Principles of Insect Pest Management.
Insect Ecology.
Principles of Biological Control.
Host Plant Resistance.

Research Projects

Extensive research programs on integrated control of the pests of cotton, sorghum, peanuts, sugarcane, tree crops and corn are being conducted by this university.

Utah State University
College of Science
Professor of Entomology: Dr. Donald W. Davis
Logan, Utah 84322

Course Offerings

Undergraduate

Utah State offers a Bachelor of Science degree in pest management. Specific courses in pest management are:
Concepts of Pest Management.

Field Experience in Pest Management.
Biological Control.

Research Projects

"Pilot Pest Management Program on Forage Alfalfa."
"Environmental Improvement through Biological Control and Pest Management."
There are also three projects, Fruit, Alfalfa, and Range Insects, which are multi-faceted studies including control of all types and pest biology.
This program is oriented toward

the practical application of biological and agricultural principles including: identification of both pest and beneficial organisms, environmental conditions in relation to pest development, sampling, control methods, economic evaluation and decision making. Students with this training are prepared to be advisors for control problems associated with agricultural, public health, forest, recreational household and urban pests.

Virginia Polytechnic Institute and State University
Department Chairman of Entomology:
    Dr. James McD. Grayson
Blacksburg, Virginia 24061

Course Offerings

This Department of Entomology has incorporated the concepts of bio-

A sample of the insect population in this soybean field is collected from a rotary trap to predict the level of parasitism at any stage of soybean growth. These predictions are based on trapping records.
(Agricultural Research, USDA)

logical control and Integrated Pest Management in their teaching, research and extension programs. Most of their formal courses contribute towards training people in control of pests through pest management programs.

Undergraduate
Insect-Pest Management.
Integrated Plant Pest Management.

Graduate
Arthropod-Pest Management.
Biological Control in Relation to Arthropods.

Research Projects

Development of Pest Management Systems for Major Arthropod Pests of Peanuts and Soybeans.
Seasonal Development and Management of Insects and Mites Affecting the Production of Stone Fruits.
Southern Pine Beetle Management.
The Principles, Strategies and Tactics of Pest Population Regulation and Control in the Alfalfa Ecosystem.
Biology and Pest Management of Tobacco Insects.
Bionomics and Population Management of Major Pests of Man and Domestic Animals.
Pest Management in Relation to Deciduous Fruit Trees.
Biocontrol of Insect and Weed Pests of Forage and Field Crops in Virginia.

Washington State University
Chairman of Entomology Department:
    Dr. R.F. Harwood
Pullman, Washington 99163

Course Offerings

A full course of study in pest management is being organized for the Fall semester of 1978. Courses in this program will include weed control, insect control, disease control, general pest control, and ecological aspects of the environment and pest control.

Research Projects

Two Integrated Pest Management Projects are being operated by Washington State University in the field:

a program in tree fruit pests and the management of these pests; and a project on IPM of alfalfa seed pests.

The objective of the tree fruit project is the improvement of timing sprays where they are required and the reduction of sprays where they are not needed for control of codling moth, mites, aphids and other insects involved in the tree fruit program. The stress of the seed program is on the use of beneficial insects in the control of lygus bugs and some of the other major insect pest species encountered in alfalfa seed fields. The seed project covers a four-state area: Washington, Idaho, Oregon and

Nevada. An attempt is being made in both programs to make better use of pesticides and reduce the use of pesticides as much as possible.

West Virginia University
Division of Plant Sciences
Assistant Professor of Entomology: Dr.
  James W. Amrine
Morgantown, West Virginia 26506

Course Offerings

Undergraduate

Insect Pests in the Agroecosystem. Presents information on the identification of insect pests from feeding and oviposition injury to plants and livestock.

Pest Management. The strategy of insect pest management using techniques such as biological control, plant resistance, cultural and mechani-

Sterile tobacco budworms may be their own enemies. Pilot studies are evaluating the effectiveness of the sterile hybrid concept for tobacco budworm control. If successful, cotton growers will have another weapon in the non-chemical warfare on cotton pests. *(Agricultural Research, USDA)*

cal methods, and timely applications of pesticides are studied.

Research Projects

Control of alfalfa insects, particularly the alfalfa weevil, is the topic of research at West Virginia University. Emphasis has been placed on IPM, specifically concerning the release and protection of several species of parasitic hymenoptera.

University of Wisconsin—Madison
Chairman of Entomology Department:
    Dr. Boush
Madison, Wisconsin 53706

Course Offerings

Undergraduate
Principles of Economic Entomology.
Biological Control of Insects.
Principles of Insect Pest Suppression.
Colloquium in Environmental Toxicology.
The Use of Chemicals in Agricultural.
Pest Control.
Insect Pathology.

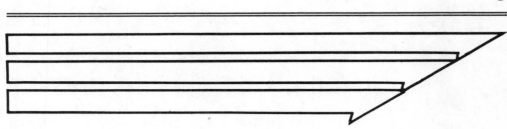

IPM for Consumers
and Policy-Makers

Chapter 7

The Consumer and Pest Management Decisions

One most significant feature of Integrated Pest Management which we believe has been completely overlooked is its potential advantages in the consumer marketplace. In other words, growers should not only look at IPM as a pest control technique, but many should assess it as a *marketing tool.*

To explore this idea, we invited Don Cunnion, a marketing specialist who had been chief of the Pennsylvania Department of Agriculture's Marketing Services, to analyze the potential.

Can an agricultural product raised under IPM conditions claim distinction in the market place? Is it possible for buyers in the marketing chain, from first handler on through to the ultimate consumer, to be persuaded that there is something special about an agricultural crop produced with a lower-than-average application of pesticides?

With bulk commodities—such as feed grains, tobacco, cotton, soybeans and sugar beets—brokers and bulk handlers will most likely *not* be very enthusiastic about an IPM identification. Most probably, crops grown by IPM methods would be grouped in with all the crops being handled. If IPM commodities were to be bought at an attractive price, brokers would be happy, but that's about as far as their enthusiasm would go.

It would appear that IPM *has its greatest marketing advantage,* other than price, *in the areas of fruits and vegetables sold for the fresh market.* That's because of the growing awareness among consumers of the hazards of chemical pesticides. More and more, consumers are voicing their concern that "too many chemicals are being used to produce our food."

How would such consumers react if offered the opportunity to buy produce grown under IPM conditions? The chances are that many would seek out such produce and might even be willing to pay a premium price for it. There are no United States market tests at the present time to bear this out, but a Swiss supermarket chain has made such an offer to its customers, with very successful results.

(Ed. note: The Migros chain of cooperative stores based in Zurich, Switzerland, placed stars over bins of fruits and vegetables to designate that they were grown without applications of chemical pesticides. The marketing manager of the stores reported that the "starred" bins invariably sold out before the conventionally-grown produce, even though the price was the same.)

Point-of-sales labeling has a direct affect on sales. It is common practice to include descriptive words on price cards—words like crispy, juicy, tasty, tender. Research has proved that these words do help motivate customers to buy. Other words such as "local" and "nearby" help convey the thought of freshness and better quality.

In Pennsylvania, market tests were done with such items as peaches, apples, potatoes and jars of honey. In each case, the use of the words "product of Pennsylvania" or something similar increased sales over identical products not so promoted. In each case all the products being offered were from Pennsylvania. Only part of them were identified as of state origin. But those few extra words doubled and tripled sales over the unidentified. To eliminate the factors of chance, locations were shifted. A more scientific study, done by the Pennsylvania State University, proved that Pennsylvania-labeled apples readily out-sold Washington State-labeled apples of identical variety, color, grade, size and price.

New York State has an on-going program to encourage the purchase of state grown foods through point-of-sale promotions, as do Texas and some other states.

But what all these programs show in the end, is that consumers do notice and pay attention to point-of-sale labeling and are motivated by it—provided, of course, a benefit is implied. (It is an old Madison Avenue maxim that to successfully promote something you must get the customer's attention, create interest and arouse desire.)

Where and how could IPM-grown produce be promoted and sold at retail? In the beginning, the best locations would be at roadside stands, public farmers' markets, papa-and-mama stores and at independently-owned supermarkets where the produce manager is permitted to buy from local growers. Food buying clubs and cooperatives, springing up across the United States, provide another outlet. The principal problem area would probably be with the brokers and wholesalers who supply corporate or volunteer chains.

Produce brokers and wholesalers deal in a highly competitive marketplace with a highly perishable product. They can be expected to purchase IPM produce on a price basis, provided the quality was equal to what they were in the habit of handling. They might, or might not, pass along the savings to the retailers, depending upon the competitive situation. Chances are they would ignore the fact that the produce was IPM grown. They wouldn't want to take the trouble to identify it as such. Once a demand was created, they'd climb on the bandwagon—but only if an adequate and continuing supply was assured.

At the retail end, some innovative market testing can be done. An IPM grower with his own direct marketing outlet, such as a roadside stand or space at a farmers' market, is in an excellent position to do market testing and start building an awareness of the virtues of IPM produce. Here's how it might be done:

Let's assume apples are being sold at a roadside stand. Several of the most popular varieties are chosen for testing. Each variety is split into two offerings of identical grade and size. One offering carries standard identification—variety name and price. The other offering carries the same identification and the same price, but is additionally identified as IPM grown.

This immediately raises the question: How is Integrated Pest Management explained in a few well-chosen words that will arouse interest and create desire? Certainly the words, "Integrated Pest Management," have minimum sales appeal—and are, in fact, meaningless to the uninitiated. A practical approach would be something like this: On the price card, or variety identification card, the words "IPM Grown" are used. Then, a highly

visible sign above the displayed fruit explains what IPM is about: "IPM stands for Integrated Pest Management. This means fruit grown with a minimum use of pesticides—only enough to assure top quality for you." Improvements would come to mind along with experience in the marketplace.

It would not be long before consumer acceptance of the IPM idea could be assessed. Hopefully, the point would soon be reached where everything sold could be advertised that it has been grown under the IPM program.

The next step would involve trying to interest a local retail grocer by using the roadside marketing experience as a sales pitch. For the initial testing with the grocer, an offer to take back anything he cannot sell under the IPM label could be

made. Suggesting that he try getting a slightly higher price for this produce is always an attractive prospect.

A successful test at the local grocery would lead to an overall marketing approach at the wholesale level. There are many ways to do this. One is to persuade a wholesale distributor to try some testing, again based on experiences with the local grocer and at the roadside market. Another avenue is for a group of IPM growers to form their own marketing cooperative, or convince their existing marketing co-op to get involved. Actually, a regular broker or wholesale distributor who became interested and turned up good test results could make arrangements with a group of individual growers to receive the needed volume on a voluntary basis. Many wholesalers or brokers would hesitate to test market something of this nature if they were not assured of an adequate supply to meet any created demand.

It appears that there are two market-ing advantages offered to IPM produce growers. One, an appeal to the expanding consumer attitude against the use, or excessive use, of chemical pesticides. Two, the appeal of price. And price can be made to work one of two ways, depending on what market testing reveals. Because of lower production costs due to savings on chemicals and their application, a cent or two could be shaved off in a tight competitive situation. Conversely, it may be possible to obtain a slightly higher price because of the IPM appeal.

It is obvious that such marketing advantages will be available to IPM innovators who get in on the ground floor. It seems very likely that the time is not far distant when Integrated Pest Management will become accepted procedure. It is at that point front-runners may find it to their advantage to abandon chemical pesticides altogether.

—Don Cunnion

Cosmetic Standards

The issue of quality standards is seen as a pivotal one in determining future use of IPM methods. The first major pest control decisions which a grower makes in a season are the contractual agreements with a wholesaler or processor regarding quality, wrote Dr. W.R.Z. Willey in his paper (cited elsewhere) on *Barriers to the Diffusion of IPM Programs in Commercial Agriculture.*

Don Cunnion makes the point that grades "were not established for the consumer, only for the wholesale buyer. Without grades, wholesalers could not buy 'sight unseen.' In contrast, in direct marketing, you don't have to worry about grades at all. The consumer does his or her own grading."

An excellent analysis of the "cosmetic standards" issue was made by Martin Brown in *Environment* magazine, (Vol. 17, No. 5, July/August, 1975) under the title, "An Orange Is An Orange." It begins:

"Special circumstances in the California orange-growing regions demonstrate how economic and marketing factors can spur massive overuse of pesticides. The California orange, revered for its lush good looks, would still be as delicious and nutritious if grown with drastically reduced pesticide treatments. Some of the oranges would have certain insect-caused skin blemishes which would not affect the quality of the fruit but against which retail customers have been conditioned by advertising. Yet to prevent these blemishes, California growers mount a multimillion-dollar campaign each year against a single insect, the citrus thrips, which normally does no more than mar the skin of the orange without in the least injuring the inside —which, presumably, is the part of the fruit which interests the consumer. So competitive and specialized is the orange industry in California that vast wastage of resources occurs as growers strive single-mindedly to produce fruit which will achieve the highest quality grade. That is the only grade which assures them profit on their crop, even though much of the fruit graded lower is of the same interior quality and would otherwise be acceptable for the fresh-fruit market.

"To see how all this has come about, let us turn our attention first to the humble thrips. The citrus thrips is a minute insect pest which does not usually reduce yields, kill trees, or lower the nutritional value or eating quality of citrus fruits. In fact, its only offense is that it changes the appearance of marketable fruit; the skin of infested fruit may be irregular or scarred to some degree. Nevertheless, this pest, a so-called cosmetic pest, is as heavily attacked in insecticide programs as are some of the more devastating enemies of agriculture."

Brown then traces the outbreak of thrips-related damage from 1908 to the present time, as listed in official federal and state agricultural department reports. "The extent to which such damage constitutes an *economic* loss to the grower," notes Brown, "however, depends not only on the severity and incidence of the scarring, but on the level and enforcement of quality standards used by packinghouses to grade the fruit."

As part of his study which Brown did while a graduate student at the University of California and which was sponsored by the EPA, he and his coworkers evaluated the contention that thrips damage has no effect on the eating quality of the fruit.

"First, two boxes of navel oranges were obtained from a local grocery store. One box carried the Sunkist label (and the fruit was marked 'Sunkist', indicating top grade). The other box was from the Valley Cove Ranch and was marked 'organic'. In both cases, the oranges sold for 29 cents per pound."

Brown found that for many criteria by which consumers measure quality, the samples from both sources were comparable. In fact, the Valley Cove (organic) oranges were superior in soluble solids-to-acid ratio, and the Sunkist sample barely met the minimum statutory standard for sugar content. Thus, in the case studied, extensive thrips activity in the untreated grove did not adversely affect the interior quality of the fruit.

Continued Brown in his study:

"Under the current quality criteria of the major citrus marketing cooperatives, a crop of oranges with 50 percent thrips damage, such as that in the Valley Cove samples, would never be marketed under the top grade. In fact, it is likely that such a crop would be completely diverted to juice. The Valley Cove oranges, which were of high interior quality, reached the market only because they were marketed by the grower. For the same oranges to reach the market by way of a major marketing cooperative, the oranges undoubtedly would have had to have been treated for thrips.

"It should be possible to educate consumers about the advantages of fruit which may have superficial blemishes but which has been grown with a minimum of pesticide use. For example, the Consumer Cooperative in Berkeley, a major retail outlet in the city that is consumer-owned and operated, recently featured lower quality oranges at 19 cents per pound as a 'superior buy' over 29-cent-per-pound Sunkist oranges. This evaluation was backed by the prestige of the store's home economists.

". . . Finally, it must be recognized that as long as the economics of California agriculture makes it necessary for a significant proportion of the citrus crop to be allocated to the processed market at a financial loss to the grower, cosmetic quality standards will continue to play a role in grading standards. Under the current system, as each grower, operating in isolation, seeks to maximize individual income by maximizing the proportion of his crop that qualifies for the fresh-fruit market and for the highest quality grade in that market, each grower must spray the entire crop for maximum cosmetic effect, regardless of how much of the aggregate crop ultimately goes to juice. (It is revealing to realize that the crop ultimately designated for juice would not need the pesticide spray treatment at all; the pesticide problem could be greatly reduced by eliminating just this unnecessary spraying.)"

Dr. Willey also concluded his report on "Barriers to the Diffusion of IPM Programs in Commercial Agriculture" with the need to focus on the regulatory arena:

"Eventually, we will have to confront the reality of the imbalances in economic power which control the information and choices available to both the growers and the consumers. As long as the sales of the pest control, food processing and food distribution industries can continue to grow at a reasonable rate and corporate equity is held intact, there is little reason to expect major changes in the direction of this momentum. . . . The problem in pest control, food processing and marketing, however, is that competitive conditions are hard to find."

Still another problem which relates to pesticide use (and non-use) has recently been revealed. Just about all major manufacturers of pesticides as well as drugs, etc. rely to a certain degree upon independent laboratories to test the safety of new products. These laboratories had been given credit for impartiality and accuracy. However, in December, 1977, one of the leading independent labs was accused of doing haphazard research and falsifying data. The accusations were made in testimony by the Food and Drug Administration and in an announcement by the EPA. Obviously, the health hazards of new pesticides are a major factor in encouraging further use of such techniques as IPM.

Testimony presented in the case against some testing laboratories revealed how a long waiting-list of products kept pressure on overworked staff to take shortcuts. A *Wall Street Journal* article included the following: "Lab technicians jotted observations on their coat sleeves, and sometimes neglected dead rats until they had decomposed too badly for an autopsy. The prevailing attitude, a lab director said, was, 'Get it out. Get it out. Shortcuts are okay.' "

The awesome power of the agrichemical giants is symbolized in the Dow Chemical publication entitled "A Case for Pesticides and Farm Chemicals " (which is available free); its purpose is "to acquaint the public with little known facts about subjects of paramount importance to everyone who eats to live or otherwise uses products of the farm and forest."

Typical of chemical industry propaganda, the slick, handsomely-printed booklet is designed to lull the consumer into a do-nothing-but-buy-Big-Brand state. Don't worry, don't complain, because you'll just raise prices and cause starvation: "Committees are banding together, proposing new laws to limit the use of needed fertilizers and ban useful pesticides, unaware that if they succeed they

Peach Growers Suggest That Those Who Don't Want Children Handle Pesticide That Can Cause Sterility

By DAVID BURNHAM
Special to The New York Times

WASHINGTON, Sept. 26—A national agricultural group has suggested that older workers who do not want children and persons who would like to get religious bans against allowed

In a letter to the Occupational Safety and Health Administration, Robert K. Phillips, executive secretary of the peach council, said: "While involuntary sterility caused by

After paying assure become cancer, rview burg,

joint emergency program to restrict the exposure of workers to DBCP and suspend its use on carrots, peanuts, tomatoes and other plants where residues consumed while eating them might cause cancer.

Because orchardists use DBCP in the soil before planting new peach trees, the pesticide is not believed to infect the peaches themselves.

In his Sept. 12 letter to Dr. Eula Bingham, head of the O.S.H.A., Mr. Phillips of the peach council said: "It appears to us that you and Labor Secretary Marshall may have overreacted, or at least that is your public posture."

is the main prob-

.A. Reports Major Differences Tests of Pesticides Rated Safe

WASHINGTON, Aug. 15 (UPI)—The Environmental Protection Agency reported today that it had found "significant" deficiencies in tests by a major laboratory that raised questions whether supposedly safe pesticides might cause cancer, birth defects, nerve damage or other problems.

Because of the discovery, the agency said 31 pesticide companies and two laboratories had been asked to audit

Chemical pollution threatens Michigan farm

By PIET BENNETT
Associated Press

GRAND RAPIDS, Mich. (AP) In 1975, George LeMunyon lost his first herd of dairy cows to chemical contamination. Using money from a court settlement

vironmental Protecion Agency (EPA) tend to agree. In fact, not until December when the EPA asked chemical to provide information on product hazards, did the

ticide: Boon and Possible Bane

By RICHARD D. LYONS
Special to The New York Times

WASHINGTON, Dec. 10—Half the chemicals being churned out by industry are pesticides, many are not only a boon to agriculture also a hazard to those who make them—and, possibly, everyone

reports of the disturbing effects poisons on workers in pesticide

First of two articles.

Virginia, Arkansas, Texas, Colorama and California have stirred pesticides could dam- believed

number of people, perhaps several hundred. But a growing number of scientists, legislators and Government officials are warning that in a decade chemicals may trigger and defects and cancer.

The core of this fear cals accumulate in the point, depending on the individual's susceptibility level that will be too even in those who do chemicals.

Pesticides already be responsible for livers, nervous system ticles of some wo

Continued on

Lake Ontario Fish Found to Contain Too Much Pesticide

The New York State Department of Environmental Conservation announced Monday that new laboratory confirmed

Consumer Group and Labor U Urge a Federal Curb on a Pest

By RICHARD D. LYONS
Special to The New York Times

WASHINGTON, Aug. 23—Use of a chemical pesticide widely employed in both household gardening and farming was attacked today by a labor union and a consumers' group because of reports linking the compound to sterility in men and cancer and birth defects in animals. Health Research Group, an arm

and animal evidence of carcinogenicity of DBCP, this test is clearly Wolfe said, adding that "Federal is needed to stop the risk to chemical workers, farm workers, and others whose jobs bring them in contact with this fumigant."

The same line of attack was A. F. Grospiron, president of the workers union, in a stant Secre onal Safety

afe milk for infants hard to find

The Washington Post

WASHINGTON — Scientists told a Senate subcommittee yesterday that it is difficult, if not impossible, to find safe milk for newborn infants.

breast milk increasingly pesticide residues and other cal contaminants that can cer and other dis while

and chairman of the Department of Human Development at Michigan State University.

"As we have contaminated our environment, we have contaminated our bodies and in doing that we have contaminated our breast milk."

Sen. Edward M. Kennedy, D-Mass., chairman of the subcommittee, said the purpose of the hearing is not to discourage b Instead for

that at le lants in May Calif

will indirectly bring about higher costs for foods, clothing and building materials.

". . . Another autumn. In the past few months, something has happened. Crops are starting to wither in the fields; chewed leaves . . . dry upon trees . . . whole flocks of chickens dying overnight. . . . There is no mystery about their deaths; for one after another many of the chemical tools that helped combat disease, weeds and insects have disappeared in a spiral of price increases, bans and government regulations, leaving not so much as the wisp of a vapor. . . . Will there be anything to harvest tomorrow . . . the day after . . . ?"

The writers and PR people at Dow then say they are by no means intending to make a parody of Rachel Carson's important work, *Silent Spring.* It's just that "this is no longer the world Carson knew."

But despite all the attempts at cover-up of pesticide hazards, many consumers in the United States are learning about potentially serious health problems from pesticides. They can understand the need and utility of pesticides, but millions of Americans know the dangers as well.

Front-page newspaper articles, as well as more detailed reports in scientific journals, report how serious health effects of pesticides have been felt by only a relatively small number of people, perhaps several hundred. Yet a growing number of scientists warn that in a decade or two, these chemicals may trigger an upsurge in birth defects and cancer. Lawsuits proliferate, with almost $200 million in damages being sought for personal injuries from pesticides.

A *New York Times* dispatch in December, 1977 reported that Dr. Bruce Ames, a biochemist at the University of California, who has an international reputation as a pioneer in cancer research, declared:

"There's no question that society is going to pay for all the pesticides that have been used in the past through increases in birth defects and cancer, but no one knows exactly what the price will be and when it will be paid."

Whenever such observations are made, the response from the pesticide industry is an automatic one that goes like this: "The cost of going without pesticides would be enormous. The National Agricultural Chemicals Association estimates that, without pesticides, the $35 billion worth of products generated by American farmers each year would be cut from 30 to 60 percent. The association, which speaks for the pesticides industry, said that without pesticides, food prices would soar and much of the $12 billion a year that this nation earns selling farm products overseas would be lost, exacerbating the country's balance of payment problems." That's quite a mouthful, but the response avoids any substantial issues, and is only designed to reassure the American public that they shouldn't be concerned with pest management strategies, that all is under control.

But such is not the case, and the American consumer must be concerned with pest management decisions. Remember, much of the increase in pesticide use is attributed to the demand by American consumers for cosmetically-perfect food.

"Many of the present government regulations (on food standards) are based solely on cosmetic appearance criteria, and have little or nothing to do with food quality," wrote H.C. Coppel and J.W. Mertins of the University of Wisconsin in their book, *Biological Insect Pest Suppression.* "Pesticide applications (to prevent harmless tiny blemishes) often terminate an otherwise successful effort at biological or integrated pest suppression. . . . If the consuming public can be taught by Madison Avenue tactics to demand cosmetically-perfect food even at the expense of the environment (and their own health), then surely they can be convinced that the quality of the environment can be improved without a loss in food quality if they will accept a good-tasting peach which is not necessarily beautiful to look at."

Clearly, education of the American public to pest management issues and cosmetic food standards is essential to help growers get off the "pesticide treadmill." The smokescreens and scare tactics put forth by the pesticide industry spokespersons are in opposition to such education. However, an increasing, and already sizable, percentage of the American public is knowledgeable about many issues affecting agriculture in general, and pest management in particular. These persons can serve as a springboard to rally others to support constructive change. The millions of Americans who garden—perhaps more than 50 percent garden with little or no use of pesticides—appreciate the value of agricultural land, and the need to take steps to preserve it and the farmers who produce food and fiber on it. Millions of Americans who are labeled as environmentalists are also convinced that important changes must be made in agricultural management. And the growing millions of Americans who belong to food buying clubs and cooperatives, and who shop at farmers' markets and buy at roadside stands, provide an increasing market for growers who use minimum amounts of pesticides—or no pesticides—to produce their crops.

In the past ten years, the market for organically-grown foods and for foods without artificial preservatives and flavors, has expanded significantly. It would seem useful to develop an understanding of the elements of that marketplace, in order to better understand the potential marketplace for IPM produced foods.

In 1970, E.B. Weiss, vice president, creative merchandising, of the advertising agency of Doyle Dane Bernbach, described the "enormously powerful trends in our society" that were increasing the natural foods market this way:

1. Higher education for more people, which makes consumers more aware of the dubious merits of some of modern chemistry's contributions to the quality (not to say safety) of certain foods.
2. The tremendous increase in disturbing news reports on additives, on DDT levels in certain foods.

(Ed. note: Today, such concerns would relate to DBCP, PCB, et al.)

3. The growing involvement of medical science in analysis of food. Medical scientists, for example, now report that certain additives which may be quite harmless individually may prove deleterious in combination.
4. The remarkable surge of public concern with respect to our deteriorating ecology. These environmental factors often cast grave doubts on some sources of food.
5. The consumerism movement.
6. And, the new food attitudes of the young people which, in turn, are part of their new concepts of living as well as their growing awareness of the questions being raised about some aspects of modern food chemistry.

Mr. Weiss made those observations almost ten years ago, and if anything, the awareness and the numbers of persons in the natural food marketplace have increased dramatically.

For millions of Americans, the term "Convenience Foods" signifies as many negative as positive qualities. Instead of time-saving advantages, this minor segment of the American marketplace translates "convenience foods" into an absence of taste, presence of preservatives, abundance of plastic wrappings and miles of transport. The young, college-educated family unit as well as the grandparents with vivid memories of food in the "Old Country" or "back on the farm" are turned off by advertising claims on TV for instant flavor and super-quick satisfaction in four-color processed foods.

The "Inconvenient Food" (food without artificial flavors and additives) market is more easily identifiable by what it *does*—rather than what it *is* in the way of age, income, education, voting preferences or mode of dress. For example, when given the opportunity, persons in the Inconvenient Food Market will have a large organic (or semi-organic) garden in their backyard, will bake their own bread, breastfeed their children, compost their garbage, walk to work when they can, carefully read ingredient labels and buy directly from local farmers by going to roadside stands, urban farm markets or joining food cooperatives and buying clubs.

The main points of the above discussion are: a small, but significant segment of the food-buying public is using its consumer purchasing power differently, and that difference offers opportunities for farmers who use minimal (or no) pesticides in producing food.

Although valid statistics are difficult to develop, probably almost ten percent of the United States consumer market is aware of the trade-off between minimum pesticide use and maximum cosmetic quality. This market segment is now being largely ignored except by a small number of organic farmers, food cooperatives and specialty stores. It is not that millions of Americans are eager to buy ugly, wormy fruits and vegetables. But an effective IPM program and a well-managed organic farm need not and do not distinguish themselves by only yielding wormy, ugly crops. There is a genuine range of consumer acceptability for "less-than-picture-perfect" food that also is grown with "less-or-no pesticides." Farmers, pest management consultants, researchers and policy-makers should consider those consumers in the aggressive adoption of IPM methods. Ignoring them will be a gross mistake!

Chapter 8

Ecosystem Management in Cities, Schools and on Highways

The state of California provides some excellent examples of how a city department, state agency and school system can utilize IPM approaches to minimize pesticide use and save money for taxpayers.

San Jose, Berkeley, Palo Alto, Modesto and Davis are California cities that have effectively used urban pest management methods to control insect pests on shade trees. Due mostly to the efforts of William and Helga Olkowski, the Public Works Department of San Jose, for example, cut city pesticide use by 90 percent.

In a paper, "Ecosystem Management:

Crew members of the Urban Integrated Control project work together with public works personnel in San Jose. Carefully timed water washing of the trees is a useful strategy for reducing honeydew accumulations and enhancing beneficial effects of the pest's natural enemies.
(California Agriculture)

A Framework for Urban Pest Control," published in *BioScience* (June, 1976), the Olkowskis and co-authors Robert van den Bosch and R. Hom explained the concept which was first developed for Berkeley's 30,000 street trees (123 species). Key components in the program were a public education effort and the importation of host-specific parasitic Hymenoptera to control various species of aphids.

The paper stresses the tremendous amount of pesticides used in urban areas and how little assessment of the hazards is made. The authors review specific parasites found to be successful against aphids on various shade trees, like the linden, elm and English oak. They then offer the following generalizations:

"For each dollar invested in classical biological control, 30 dollars have been returned in savings. The research and development costs for one synthetic chemical pesticide are about $5 million or more, whereas the average cost per entomophagous species introduced successfully ranges two to four orders of magnitude lower (some have been obtained for less than a few hundred dollars).

"The classical biological control technique is ideally suited to urban insect population management. High human densities require the employment of control methods that are of low risk to mammals and other animals. The high degree of insect control demanded by agriculture is not generally necessary for shade trees. Rarely are such insect pests of economic or medical importance. Usually they can best be described as mild nuisances, their effect being primarily aesthetic, or one of repugnance to a populace afraid of insects. Any degree of population control, which makes the insects and their activities a little less visible, may be all that is needed to appease citizen demands for pesticide use.

"Biological control is well suited to urban areas because much of the vegetation is perennial. Such vegetation is not subject to the disruptive practices (i.e., harvesting, irrigation, sanitation, plowing, burning and rotation) which affect agricultural plantings. Furthermore, the diversity of urban vegetation, at times involving hundreds of species, must surely contribute to the stability of insect communities."

Since 1971, the Olkowskis' Urban IPM Project has taken concepts developed in agriculture, plus alternatives to chemical controls researched both within and outside of universities, and have adapted them to non-agricultural systems—i.e., cities and schools. "Our aim," they explain, "has

Tulip trees on the streets in Berkeley, California are monitored for Tulip aphids as part of the city's biological control program. *(California Agriculture)*

been to develop model programs that might be of some use to the private pest control industry, and could provide a basis for a special urban pest management consultant approach."

The success of the shade tree programs in various cities encouraged the Olkowskis to try to reach out to the schools. They decided to develop a model IPM program for an entire school district. ("By educating school age children, it might be possible to affect whole families.") Their efforts led to a program in the Palo Alto School District.

The most common pesticide materials used in the schools were chlordane for ant control, Diazinon against roaches and 2,4-D for weed management. Oak trees were also sprayed for Oak Moth control. In addition, custodians would bring in insecticides on their own, as if insect presence in the schools was a sign of personal incompetence.

During 1977, their second year of working with the school district, the Olkowskis achieved the following developments: "Pesticide use was confined to petroleum distillates for weed management, boric acid in the cockroach program along with minor use of a pyrethin ant and roach spray, some bleach and a great deal of soap and water."

Control efforts for ants and cockroaches also involved sealing up wall cracks with caulk or putty, eliminating food sources and paper clutter, setting traps and careful monitoring.

Future plans for the Palo Alto school district program include building IPM into the curriculum—at all levels, but especially in high school science and home economics classes. Add the Olkowskis: "Educational materials should demonstrate how human actions and aesthetic judgments are often responsible for the creation of problem situations, and provide common sense healthy management strategies that can be applied in school and home environments. A few students may be encouraged to consider careers in IPM fields."

California Transportation Department Avoids Sprays

Dan Cassidy, a landscape maintenance specialist with the state of California's Department of Transportation (Caltrans), knows what it is like to be caught in the syndrome of intensified pesticide usage.

"We were using more and more pesticide and every year the insect problems were getting worse," states Cassidy. "I was very concerned. We knew something was wrong with our increased pesticide use, but we didn't know what was wrong." The 15,750 planted acres along highways make the Department one of the largest horticulturists in the world.

University of California (Berkeley) Professor Dudley Pinnock, an insect pathologist, shared Cassidy's concern. "Unfortunately," said Dr. Pinnock, "there is

(This report on Caltrans is written by Joel Grossman.)

no magic answer for many pest problems. We can only do what is biologically possible."

Landscape specialist Cassidy and IPM specialist Pinnock combined their talents to develop a control program for insect pests of the California highway landscape. According to Cassidy, "The long-established goal of the California Department of Highways has been to maintain its landscapes as economically as possible and with minimal use of chemical pesticides."

Plants growing alongside the highways are often highly stressed before the insect pests arrive. Road building operations frequently remove much of the good topsoil and road grime may coat the leaves.

Leaf eating worms, such as the California oakworm, fruit tree leafroller, tent caterpillar and the red-humped caterpillar, on the trees were costing the highway maintenance people the most time and money in control costs. The common cotton (melon) aphid and the citrus aphid were major pests on shrubs, such as pyracantha. Anywhere from one to several spring sprays were being used for aphid control. Additional summer sprays followed for worm control. Pesticides such as diazinon, carbaryl (Sevin), malathion and methoxychlor were most commonly used.

Dr. Pinnock suggested that the bacterium *Bacillus thuringiensis* (tradenames: Dipel, Thuricide and Biotrol) be used for worm control. In other words, microbial control would substitute for chemical control in this situation.

Landscape specialist Cassidy said, "We didn't have any success when we first used *Bacillus thuringiensis*. So we told Dr. Pinnock that it didn't work. Nevertheless, we agreed to let him personally test it

using our equipment." The Professor demonstrated the microbe's value in worm control and led Cassidy to conclude, "We just didn't know how to use it at first. *Bacillus thuringiensis* provided better worm control than the chemicals." Also, native beneficial insects were seen again.

"This demonstrated to the highway people that there was an alternative strategy in insect control," said Dr. Pinnock. "Since then, we have gone past the stage of using *Bacillus thuringiensis* as a substitute insecticide. We now adjust the application rate to provide the exact worm kill that is ecologically desirable."

If all the red-humped caterpillars on a tree were killed, then there would be no food source to sustain beneficial *Apanteles* and *Hyposoter* parasites that provide natural biological control. "So, what we want to do is reduce the worm population to a level which will sustain the beneficials and not cause economic damage to the plant," said Dr. Pinnock."

The number of pest control operations for worms was reduced considerably —often from half a dozen pesticide applications to one or even no microbial treatments. Thus, workers were freed for other tasks and the need for more spray rigs was reduced.

Aphids feeding on the landscape shrubs were sometimes being sprayed with pesticides every spring. Beneficial insects, feeding on the nectar of flowering shrubs, will move onto other plants in the landscape to search for pest species as prey. However, the California highway landscape lost this valuable source of beneficial insects when chemical insecticides were applied to control aphids, because the chemicals were not specific to aphids. The beneficial insects were killed along with the aphids. In the absence of beneficial insects, some pests quickly rebounded.

Soapy water had been used to wash off and drown stem aphids back in the 1800s. But in the 1970's, soap was an unregistered material. Cassidy said, "The laws and regulations prevented us from using the soap even though it was commonly used and recommended by the Department of Food and Agriculture 50 years ago."

An organic biodegradable soap with a coconut oil base passed the requirements for registration as an insecticide in California. A dilute spray (one to two quarts of soap per 100 gallons of water) is applied at a carefully controlled rate and pressure. The soap spray leaves about 25 percent of the stem aphids on the plants as a food source for the beneficial insects and washes off roadside dirt and grime.

The soap is sold commercially in California as Acco Highway Plant Spray. The manufacturer (Acme Chemical Company, Whittier, California) reports, "We are gaining other users such as counties, cities and the State of California's Parks and Recreation Department. The newest interest is from the State of Texas where experimental work similar to Dr. Pinnock's is now going on."

Annual training academies teach personnel how to apply IPM knowledge, such as beneficial insect identification and what constitutes an economic level of a particular pest. The workers go back to their home districts and spread the information statewide. Workers recently learned how to identify two species of moth which have been imported as part of a current research effort on biological control of tumbleweed. Tumbleweed control costs Caltrans over one million dollars per year—not to mention the costs of farmers, irrigation districts and others affected by tumbleweed.

Integrated Pest Management is more than just a pest control strategy for farmers. It is a network in which city administrators, consumers, policy-makers and students can participate by voicing their concerns about pesticides and suggesting alternative controls, thereby creating a safer environment for people on farms, in cities, villages and suburbia throughout the United States.

4

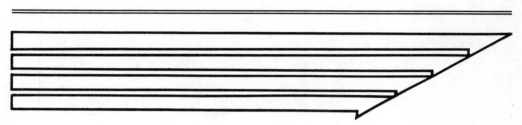

IPM Research-Past,
Present and Future

Chapter 9

IPM in Laboratories and in Field Trials

Useful and imaginative research concerning integrated controls of weeds, insect pests and plant diseases is being conducted at agricultural research centers, universities, experiment stations and on individual farms throughout the world. These large-scale studies also measure the economic consequences of altering present pest control practices and substituting alternative strategies. Several of these pilot studies are discussed in this chapter by Dr. Waldemar Klassen, a staff scientist in Pest Management with the USDA in Beltsville, Maryland.

The range of projects verifies the potential of IPM for all food and fiber production.

By no means do all of the biological control examples come from modern research. A survey of ancient writings showed that Pliny the Elder said that a crayfish placed in the middle of a garden would keep caterpillars away. As a weed control, Cato recommended the application to the soil of amurea, residue obtained when the oil is drained from crushed olives. Mixing vegetable seeds with cypress leaves before sowing was said to discourage maggots from eating the plant roots later.

A survey of ancient agricultural practices was done by Allan E. Smith, a chemist with the Regina Agricultural Research Station in Saskatchewan and his wife, Dr. Diane M. Secoy, a biology professor.

For the treatment of fungal diseases of plants, such as blight and mildew, two main methods were employed. Democritus advised that, as a protective measure, all seeds be soaked in leek juice before planting. Several others recommended that smoke from burning chaff, crabs, dung or ox horn be wafted over threatened or suffering plants. Extracts of lupin flowers or wild cucumber were widely used against a variety of pests.

The selective placement of plants that repel pests in rows alongside crop plants and vines was also recommended by the classical writers. Such plants, reported to kill or repel insects, included bay, cedar, cumin, fig, garlic, ivy and pomegranate.

In the final years of the twentieth century, our scientists and farmers are delving deeper into the intricacies of pest management.

Apple Orchards

In 1977, approximately 234 tons of pesticides were applied to apple orchards in Massachusetts alone. The pesticides cost growers $1,380,000. The apple crop, grown on 338,000 trees on 7,721 acres, earned $10,700,000.

The typical Massachusetts apple grower, explains entomologist Ronald J. Prokopy, applies 9 to 11 insecticide sprays, two to four miticide sprays, 12 to 14 fungicide sprays and zero to one herbicide spray. According to Prokopy, a professor at the University of Massachusetts, the major materials used are:

Insecticides—Guthion, Imidan, Zolone, Sevin;
Aphicides —Systox, Thiodan, Phosphamidon;
Miticides —Plictran, Omite, oil;
Fungicides —Captan, Benlate, Glyodin, Thiram, Cyprex, Polyram, Dikar, Difolitan; and
Herbicides —Paraquat, Simizin, Glyphosate.

When apple growers use these pesticides, they do so as insurance against a number of pests: apple maggot, plum curculio, tarnished plant bug, European apple sawfly, codling moth, spider mites (European red mites and two-spotted mites), apple aphids, various leafrollers and San Jose Scale. The major problems (in approximate order of importance) are: apple scab, black rot, powdery mildew, cedar apple rust and fire blight.

While these pests are capable of devastating fruit harvests, Dr. Prokopy and his colleagues discovered after a series of careful observations that pesticides are by no means the best way to protect fruit. The reason is that pesticides also kill beneficial predators and parasites and thereby induce build-up of secondary pests, particularly spider mites and apple aphids. Following is a description of their findings by Prokopy.

"We sampled numerous commercial and abandoned orchards in Massachusetts in 1976 and 1977 and found 23 species of mite predators and at least six species of apple aphid predators. The most abundant mite predator was the phytoseiid mite, *Amblyseius fallacis. Stethorus punctum* and *Typhlodromus pyri,* the principal mite predators in commercial apple orchards in Pennsylvania and western New York, respectively, were not found in any of the orchards sampled. The most abundant aphid predator was the midge *Aphidoletes aphidimyza.*

"In commercial and abandoned orchards where these two species were sufficiently abundant, populations of spider mites and summer apple aphids were low, never reaching the economic injury level. However, field plot and/or laboratory studies revealed that the insecticides Zolone, Sevin and Systox, the fungicide Glyodin, and the herbicide Glyphosate were all highly toxic to the two strains of *A. fallacis* tested. Also, the fungicide Benlate interfered with the reproductive capability of these two strains. Those commercial orchards' samples which utilized some or all of these six materials had few or no *A. fallacis,* had spider mite populations which exceeded the economic injury level, and required more miticide applications than orchards not receiving these materials."

Dr. Prokopy and his colleagues have undertaken research which stresses a multi-faceted, non-insecticidal approach to managing mid- and late-season pests in New England fruit orchards (previously referred to in Chapter 2). His non-spraying weapons include spheres painted dark red resembling nearly ripe apple fruit which attract and trap apple maggot flies (these spheres can be treated with volatile components of apple fruit and a male sex pheromone). The spheres have proven highly effective in directly capturing and controlling the apple maggot in small orchards. In addition, the sticky-coated traps, hung from branches, have proven more effective than any other method for accurately monitoring population levels of plant bugs, sawflies and apple maggot flies.

Developing methods for monitoring secondary arthropod pests such as spider mites, apple aphids, leafrollers and scale insects which primarily affect tree foliage and limbs is another research area being explored. Generally, growers are advised to spray when seven mites per leaf are found in apple orchards. But Prokopy's research will consider whether sprays are actually necessary at that level if there are sufficient *A. fallacis,* the principal predator of spider mites. Similar programs are being devised for "spray or no spray" decisions for other pests.

Previous research showed that *alternate* middle spraying of apple trees was just as effective as *every* middle spraying against key pests. "In 1977," says Prokopy, "we conducted an in-depth cost-benefit analysis in one of three orchards. The results re-

vealed a 50 percent reduction in pesticide usage and a greater *net* profit of $66.00 *per acre* in the alternate middle block compared with the every middle block."

In the past 15 years, spray schedules and pesticide usage in commercial orchards in Massachusetts have remained at the same high levels. Prokopy and his fellow entomologists at the University are confident that farmers are ready to adopt new techniques. A meeting in early 1978 with leaders of the Massachusetts Fruit Growers Association revealed that members are "fully prepared to provide the necessary equipment and supplies and to hire the necessary scouts to continue and expand the program." The required number of growers needed to participate in the research project has been fulfilled.

Wild Oats and Mosquitoes

Wild oats thrive on more than 28 million acres in the United States, causing annual crop yield losses between $300 and $500 million. It is estimated that about 90 percent of the small grain acreage in North Dakota alone is infested with wild oats, resulting in losses from $150 to $250 million annually. The Cooperative Extension Service in North Dakota has organized a project called the Integrated Wild Oat Pest Management Project whose purpose is to reduce wild oat infestations to levels which would not lower crop yields. Control is based upon careful decision-making in terms of the immediate and long-range cost benefits of an integrated control system by trained people working with farmer-cooperators. Professional advisory services are used so that the total farm management system, including pest densities and crop, soil and weather conditions, is considered when making decisions about pest control procedures. Monitoring is playing an important role in this project. Scouts are employed during the summer months to inspect the fields for pests including insects, weeds and diseases.

Wild oat seed is separated from soil samples from each farm and tested for germination to give a general indication of the severity of infestation. Seeds that do not germinate under optimum conditions will be treated with gibberellic acid to break dormancy. By this technique the number of viable, dormant and dead seeds per acre can be determined, a useful technique for measuring the potential infestation and providing a base for determining level of control.

Dr. Robert van den Bosch of the University of California's Division of Biological Control describes California's Marin County's integrated control of mosquitoes project:

"Dr. Allen D. Telford and his colleagues in the Marin County Mosquito Abatement District have developed a mosquito integrated control program which ranks with the most sophisticated in the country. The program which involves population monitoring, selective pesticide usage and breeding source management has resulted in a more than 90 percent reduction in pesticide spraying while effecting an overall reduction in the mosquito problem. The most striking element of this program has been the management of mosquitoes in the 2,-000 acre Petaluma marsh. At one time this wetland was a major mosquito producer, contributing to both urban and livestock problems. As a consequence, most of the marsh was sprayed with deadly parathion by aircraft five times a year.

"Dr. Telford and his colleagues shrewdly deduced that the mosquito source was not the marsh's maze of sloughs and channels, which are subject to tidal flushing, but instead "pot holes" in part created by dummy bombs dropped during World War II when the area was a practice bombing range. Other human activities created most of the remaining pot holes. The pot holes were not subject to tidal flushing; after flood tides many retained water and became ideal mosquito breeding sites.

"The Marin county entomologists acquired a ditching machine, and developed a pot hole drainage system which permits tidal flushing. The program has been so successful that only a few, as yet undrained, holes require hand spraying. Today, no aircraft drone over Petaluma marsh, dribbling their lethal organophosphate excreta onto the teeming life system. Yet, the mosquito problem has disappeared from nearby communities, and dairymen operating adjacent to the marsh have told Telford that their herds are free of tormenting mosquito swarms for the first time in memory.

"Thus, through an imaginative integrated control effort, California's Marin County and neighboring Sonoma County have realized substantial ecological, economic and sociological benefit."

Exercises In Applied Ecology

The Huffaker Project, coordinated by Carl Huffaker of the University of California at Berkeley, reflects many of the new trends in IPM research. This project instructs farmers how to use chemical insecticides discriminately, and also illustrates new ways of conducting agricultural research. Agronomists, ecologists, economists, entomologists, plant pathologists and systems analysts at 18 universities were involved in the project. The

Huffaker Project emphasized the *use of computer technology and systems analysis* to construct models of crop ecosystems that can provide farmers with reliable information on controlling insect pests and other aspects of crop management. This $14 million research project, funded by the NSF, EPA and the USDA, began in March, 1972 and was completed in March, 1978. The crops studied in this project included alfalfa, citrus, fruits, cotton, pome and stone fruits, and also pine forests. Models describing each ecosystem were at the nucleus for each crop subgroup. The ultimate goal concerned the use of the ecosystems in conjunction with current data from field and weather monitoring to predict when-or-if-an insect population will exceed the economic threshold. Jean L. Marx in *Science* (March 4, 1977) called the Huffaker project an exercise in applied ecology.

The "dial-a-bug" service for orchard owners in Michigan illustrates how the models can be applied. A result of the pome and stone fruit subproject, this service provides owners with information about how to manage their crops and protect them from pests such as the codling moth and spider mite. Throughout Michigan, daily information is collected about the weather, insect populations, etc. in each of 27 regions, and is fed into computer terminals for analysis by the central computer. The computer utilizes the model to derive information about crop management needs, including whether or not to spray for certain pests. The agricultural extension agents then receive this information and record the appropriate telephone message for each region.

Michigan investigators have compared pesticide use and crop yields in 200 orchards whose owners use the message system with those in control orchards and have concluded that the system permitted a reduction in insecticide applications of about 30 percent. One of the scientists with the project found that the savings on insecticide expenditures more than make up for the cost of monitoring the orchards

This sex pheromone trap, which attracts corn rootworm beetles, can save farmers money directly by assisting them in the identification of fields that do not require chemical treatments for rootworm.
(*Agricultural Research, USDA*)

and running the computer system. Similar programs are being conducted in California, New York, Pennsylvania and Washington, where reductions in insecticide applications are also being achieved.

Breeding insect-resistant plant strains is a major aspect of the cotton, soybean and alfalfa subprojects. Another line of investigation for controlling pests, pursued by the research scientists involved in the Huffaker project, dealt with the use of pest parasites or predators, some of which must be imported since the target species are not native to the United States. As is evident, this large-scale research project provided a wealth of knowledge concerning effective implementation of Integrated Pest Management in crop ecosystems where pests, their parasites, and predators, the weather and all of these factors influence each other.

Ecology is also being applied in onion patches for pest and weed control. Entomologists at Michigan State University (MSU) are studying the complex interrelationships between plant and animal species in the onion fields so they can decipher which pests need to be controlled. Dean Haynes, a MSU entomologist and population dynamics specialist, divides the plant and animal species in onion patches into five "orders." The first order includes those that attack the onion directly, such as the onion maggot, onion thrips, nematodes and a disease named leaf blight.

Each of these first-order members has other species that attack it, these being their enemies, and so on through five levels of rankings. This information has been computerized and as further research refines the workings of the system, more management control of the system can be achieved.

An example of this multi-factor control involves the destructive onion maggots, which damage onions by feeding on the lower part of the stem or bulbs; one maggot is capable of destroying many seedlings. The onion maggot proceeds through several stages, from egg to larval to pupae, emerging as an adult fly, which lays its eggs on onions, leading to the next generation. Until recently, these pests were controlled by chlorinated and organosphosphate-type compounds. However, onion maggots have developed resistance to these pesticides. Entomologists at MSU found that a fungus, called *Entomophthora muscae,* a pathogen growing in onion patches, causes abdominal bloating in the onion maggot. According to Ray Carruthers, a doctoral student in entomology at MSU, "the disease grows internally, and the fly's vital organs are dissolved. It becomes sluggish and can't fly very well." The fungus reproduces by laying spores inside the maggot's abdomen. In order to reach the outside, these spores burst through the abdominal wall of the maggot.

Carruthers has confidence in this fungus, "We have great hopes for understanding and using this disease as a control agent of the onion maggot in the future."

Although pesticides are used minimally in IPM systems, it is of utmost importance to apply them efficiently and carefully. An EPA report, "A Study of the Efficiency of the Use of Pesticides in Agriculture" by Dr. Rosmarie von Rumker and Gary L. Kelso (EPA-540/9-75-025), stated that 50 percent of insecticides and miticides used on corn and sorghum and 20 to 30 percent of the pesticides used on apples are unnecessary. *Overapplication,*

Typical Losses from Aerial Foliar Insecticide Application

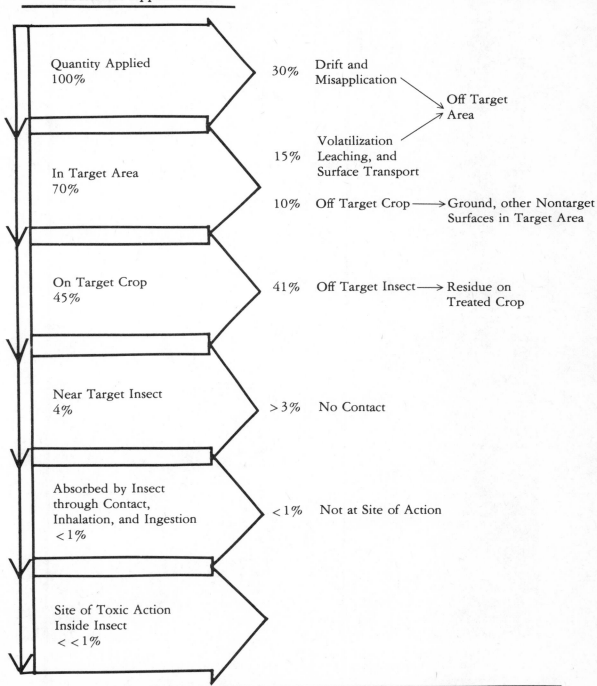

Quantity Applied
100%

30%   Drift and
      Misapplication

                          Off Target
                          Area

In Target Area
70%

15%   Volatilization
      Leaching, and
      Surface Transport

10%   Off Target Crop ⟶ Ground, other Nontarget
                        Surfaces in Target Area

On Target Crop
45%

41%   Off Target Insect ⟶ Residue on
                          Treated Crop

Near Target Insect
4%

>3%   No Contact

Absorbed by Insect
through Contact,
Inhalation, and Ingestion
<1%

<1%   Not at Site of Action

Site of Toxic Action
Inside Insect
<<1%

Percentage distribution figures do not represent one specific case, but are composited from several different field studies and assumptions on the degree of ground cover and the density of target insects on the target crop.

Source: A Study of the Efficiency of the use of Pesticides in Agriculture by Rosmarie von Rumker & Gary Kelso (for EPA), July 1975.

nonuniform distribution and drift are the major causes of inefficient pesticide use. Examples of pesticide abuse and inefficiency include use of pesticides in the absence of an established need to control or suppress the target pests, use of a pesticide at a rate of application higher than necessary for the intended pest control or suppression purpose, and use of pesticides not registered or not suitable for the intended purpose. The report suggested the reduction of this pesticide inefficiency in agriculture by promoting development of application methods and equipment that will reduce pesticide wastes and losses during application, and the promotion of soil conservation practices that will reduce post-application pesticide losses due to runoff and erosion.

Close-up of a ULV (Ultra Low Volume) sprayer. This sprayer can be pulled by a 15 to 18 hp tractor, thereby reducing fuel needs and soil compaction.
(Agricultural Research, USDA)

A fine example of how sprays can be more effectively applied is offered by the ULV (Ultra Low Volume) sprayer, made by an English company, Micron Sprayers, which sells them primarily in Third World countries. The company's original aim was to economize in pesticide use to reduce spraying costs for poor farmers, but the firm achieved more: a simple sprayer that is low-cost and markets at a fraction of the price of a conventional knapsack or boom sprayer. It has a reservoir at one end of the carrying handle with electric torch batteries inside the handle and at the working end, a tiny 12v electric motor that operates a spinning disc.

The pesticide is fed onto the disc and emerges from it as a fine spray in which all the droplets are of the same size. This allows a mist to enter a crop almost horizontally rather than vertically, so that a much more even application is achieved. Depending upon the particular use, only one to ten percent of the pesticide active ingredient is used compared to other sprayers. No pesticide drips to the ground, and control is at least as good as control with other sprayers. The application is so precise that it becomes possible to spray only adult insects and only when an infestation is confirmed, so that a ULV sprayer can be incorporated quite effectively into an IPM program.

The Texas Agricultural Experiment Station has warned farmers that *certain pesticides can cause major pest problems.* In the January, 1978 issue of *Agricultural Research in Texas,* growers are advised to use all available knowledge in order to achieve maximum utilization of beneficial insects and selectivity of pesticides so that usage of pesticides and environmental pollution are decreased. Examples of how certain pesticides may cause major pest problems through the removal of natural enemies of minor pests in citrus include:

1. citrus mealybugs become a major problem when phosphate pesticides are applied on certain mites and scale insects;
2. Texas citrus spider mites and four other scales (chaff, California red, purple and Florida red) appeared when Sevin was used on brown soft scale.

The remainder of this chapter is divided into three sections. The first section deals with beneficial insects; controls such as sex attractants, plant breeding, soil fertility and interplanting as IPM techniques are presented in the second section; and new directions in weed control are discussed in the last section. These reports on recent developments in the non-chemical forms of pest management document how IPM is rapidly becoming a viable strategy supported by agricultural experts, farmers and legislators for the provision of high crop yields and a safe environment.

Beneficial Insects

Beneficial insects, when allowed to proliferate, are effective control agents of insect pests. An overview of predators and parasites on which research and projects have recently been conducted (1977–1978) is presented here. (Please refer to

Appendix 1 on Biological Pest Control Agents for a complete listing.)

Caterpillars

The state of Oregon is employing caterpillars to attack the noxious weed, tansy ragwort, which is infesting 4.3 million acres of land in Oregon. Tansy ragwort has also been found to be poisonous when consumed as a steady diet by cattle. Charles Phillips, a biologist with the Siuslaw National Forest Service, is in charge of the caterpillar infantry, aided by youths in the Youth Conservation Corps and National Recreation Administration programs. Phillips directs the collection and transference of cinnabar larvae colonies to districts of Siuslaw National Forest lands. Thirty of 40 larvae will completely defoliate a tansy plant on cut-over forest land. The strength of this control is dependent upon an entire colony of 1,000 to 1,500 caterpillars all chewing on a tansy patch. Although Phillips has no illusions about wiping out all of the tansy ragwort in western Oregon, he believes that this program may reduce the weed to a tolerable level, perhaps 90 percent. (Source: *Western Farmer,* Vol. 55, No. 1, November-December, 1977)

Dung Beetles

A quarter of a million imported beetles are working underground to help California cattlemen and dairymen overcome a triple-threat pest problem. Raised and released by University of California scientists at Davis and Berkeley, these dung-burying beetles have been distributed in 14 California counties. Their job is to remove breeding areas of the horn fly and the face fly and to consume the slabs of dung which smother growth of new grass needed for cattle to graze on. The project, launched on a large scale in 1976, could have a big impact on beef cattle, according to Dr. John R. Anderson, professor of entomology at UC Berkeley, who is leading the project with Extension Parasitologist Edmond C. Loomis, UC Davis.

Dried-out dung pads can smother vegetation for up to four years, Loomis and Anderson point out, thus depriving animals of range feed, which is the diet for about 65 percent of the nearly 3,000,000 beef cattle produced annually in California. After four years, the resulting losses on a 5,000-acre ranch can amount to $3,000. Loomis and Anderson believe that the imported dung beetles could markedly increase forage production and reduce numbers of parasitic worms as well as horn and face flies by burying only 25 to 50 percent of existing dung pads. Four species

More than a quarter million of these imported dung-burying beetles are beginning to eliminate grass-smothering dung pads on California rangelands. They're distributed by the University of California.
(Photo by Tracy Borland, UC Davis)

of beetles are being mass-reared at UC Davis. The scientists released 7,000 beetles in 1975, 38,000 in 1976 and about 250,000 in 1977. Another 250,000 were released in the spring and summer of 1978. "These beetles will constitute a permanent, self-perpetuating type of pest management," said Loomis and Anderson. Insecticide costs for horn and face fly control are about $1,500,000 yearly.

Similar projects in Australia and Texas have been successful, the insect experts reported, but it took several years for the beetle populations to "explode" and disperse widely.

European Parasitic Wasps

The European parasitic wasps are being used in Ohio to control the alfalfa weevil *(Hypera postica),* one of the most destructive pests of alfalfa. Ohio has been releasing these parasites in order to establish strong battalions to fight the alfalfa weevil population, and spreading of this parasite has occurred naturally at the rate of 20 to 25 miles per year.

The female parasitic wasp succeeds by disrupting the reproductive processes of the weevils and eventually causing their death. After determining that the weevil is a suitable host for her progeny, the female wasp thrusts her ovipositor into the body cavity of the weevil and deposits a single egg; a larva emerges from the egg and feeds inside the weevil. The combination of sterilization and death of the adult alfalfa weevil is what makes this parasite so effective. Data from the eastern United States have indicated that 80 percent of the female weevils can be sterilized after only 50 percent of this parasite's eggs are deposited in the weevils. Optimum rates of

parasitism would give over a 50 percent reduction in alfalfa weevil larval population.

Cereal-leaf beetles damage wheat, oats and barley by chewing out the chlorophyll-containing cells of the leaves, giving the damaged or dying crops a white frosted appearance. The wasps eat their way out and thus devastate the beetle. Dr. Harold F. Willson of the New York State Cooperative Extension Service organized a campaign against the cereal-leaf beetle in New York by using European predatory wasps. New York state and federal agricultural agencies, concerned about dangerous pesticide utilization, became informed about the success of predatory wasps in Europe where the beetle population is no longer considered perilous due to their native wasps. According to Peter S. Doth, a New York state horticultural inspector, biological controls usually only must be repeated until a parasite population becomes established. Paul Lane, director of the Division of Plant Industry of the New York state Department of Agriculture, advocates biological controls over pesticides because they are more economical.

Pennsylvania and West Virginia State Agricultural Departments and Extension Services are also using European predatory wasps to reduce the cereal-leaf beetle population in oat fields.

In 1974, Maryland farmers were spending $500,000 annually for pesticide foliar applications to protect their bean crops from the Mexican Bean beetle. The Maryland Department of Agriculture reported that in 1976, these farmers maintained the same high level of production while reducing foliar application costs to approximately $50,000. This accomplishment is directly attributable to a highly successful research program at the University of Maryland Agricultural Experiment Station.

The University's team of entomologists demonstrated that a tiny wasp, harmless to humans and bearing the name, *Pediobius foveolatus,* could control the Mexican bean beetle quickly and dramatically. The insect, native to India, preys on the plant-eating beetle by depositing its eggs on the larva. When the eggs hatch, the larva of the wasp destroys the beetle larva.

Fire Ants

The fire ant is usually considered a pest to be eradicated by massive government spraying programs. However, in an announcement from the Texas Agricultural Experiment Station, the fire ant is praised because it kills so many harmful cotton insects. Texas entomologists, led by Dr. Winfield Sterling, have found that the red imported fire ant is one of the best predators of boll weevils and bollworms. Fire ants and other predators were able to eat an average of 93 percent of the bollworm eggs on cotton. In a dryland cotton field containing large numbers of fire ants where no insecticides were used for the control of any pest, 1,810 pounds of seed cotton were produced, compared to only 577 pounds in adjacent fields treated with insecticides for cotton fleahoppers, boll weevils and bollworm-budworms. "Insecticides failed to control the budworm," continues the Texas Research Report, "but where insecticides were not used, fire ants and other predators were effective, not only in preventing damage and saving cost of insecticides but also in considerably increasing yields." Instead of trying to kill it with poisons, the method of choice a few years back, Dr. Sterling and colleagues advise growers to "leave it alone in cotton fields where it works for free."

Ladybird Beetles

Pennsylvania State University entomologists have found that the ladybird beetle reduces the amount of chemical spray needed to control European red mites in orchards by 80 percent. "In many cases, the ladybird beetles have eliminated miticide sprays formerly used to control potentially harmful mite populations," explained Larry Hull, graduate assistant in entomology at Penn State. Called *Stethorous punctum,* they have the ability to search out isolated pockets of mites. Success of the integrated pest control systems, according to Hull, depends upon applying insecticides tolerated by the ladybird beetles. As the mites increase, the female ladybird beetle lays eggs which hatch into larvae that also feed on the mites. Once established in an apple orchard, the ladybird beetle adults move around in response to the highest mite populations. The most mites are usually found where the leaves have not received heavy applications of chemical sprays.

Natural Controls of Green Peach Aphids

A primary objective of a Pennsylvania State University experiment was to determine the influence of green peach aphid populations on potato yields and the influence of natural enemies on aphid abundance. When the carbaryl spray was omitted, savings resulted from the conservation of beneficial insects, which helped to reduce aphid pest population levels. The integrated pest program was found to be more economically feasible. According to the researchers, the integrated approach resulted in a $24.00 per acre savings in cost of systemic insecticides, plus the savings from not needing additional foliar application and an additional $24.00 per acre from increased yield. (Source: *Science in Agriculture*, Volume XXIV, No. 3, Spring 1977, a publication of Penn State's College of Agriculture.)

Natural Controls of Soybean Insect Pests

In the summary of "Natural Biotic Agents Controlling Insect Pests of Missouri Soybeans" by C.M. Ignoffo, N.L. Matston, et al., these scientists conclude from their data that naturally occurring parasites, predators and entomopathogens efficiently suppress populations of potential insect pests in Missouri soybeans. The report continues:

"Probably this situation exists in Missouri because minimum use is made of pesticides. In most instances, soybean growers in Missouri and probably throughout the Midwest do not need to do anything! Natural biotic control factors suppress pest populations even though caterpillars are present year after year. However, we want to be able to predict incipient damaging populations if they do develop. Furthermore, we want to be able to provide growers with safe, selective agents that are compatible with the beneficial insects and diseases and have the least impact on the environment and future agricultural technology. If we judiciously select our insect control agents and use them prudently, we may never experience the problems that have developed in other crop ecosystems."

The beneficial predator Podisus attacks a cabbage looper larva; this predator also attacks the soybean podworm, the green clover worm and the soybean looper.
(Agricultural Research, USDA)

Pine Bark
Beetle Predators

Researchers at the Auburn University Agricultural Experiment Station report that the presence of bark beetle predators is directly related to the presence and activity of the bark beetles themselves. Experiments were conducted on two predators of the bark beetle: *Medetera bistriata,* one of the long-legged flies, and *Thanasimus dubius,* a checkered beetle. Given a choice between beetle-infested and uninfested pine bark, results from the investigation clearly revealed that the adults of each predator species preferred the bark beetle-infested pine material. This discovery shows that the predators of these pine bark beetles have no trouble in locating their hosts, thus enabling them to function as an effective regulating agent.

Spined Stilt Insect

A system of augmentation and conservation of natural enemies to suppress the tobacco budworm and tobacco hornworm on tobacco is being developed by North Carolina State University and the Agricultural Research Service. The pilot test, led by scientist Dr. A. Baumhover, is designed to develop the use of natural enemies in order to reduce problems such as hazard to field workers, destruction of natural enemies, residues on tobacco and resistance associated with conventional insecticides.

The pilot test is evaluating the early season release of an important egg predator, the spined stilt bug. Pest populations that nevertheless reach a threshold justifying treatment will be suppressed with *Bacillus thuringiensis,* the latter being non-toxic to beneficial insects but effective against hornworms and budworms.

Trichogramma

Trichogramma have been receiving increasing consideration recently as a biological control agent for lepidopterous pests on field crops, reports W. Joe Lewis of the Agricultural Research Service in Tifton, Georgia. These minute wasps offer a great deal of potential for the management of a number of pests.

Other than the usual climatic considerations, explains Lewis, the primary factors determining performance of *Trichogramma* include: quality of *Trichogramma* (genotypic quality and phenotypic quality), number of *Trichogramma,* release procedures and techniques, size and density of plants, density of hosts and Trichogramma behavioral factors.

For the most part, to date, applied programs with *Trichogramma* have not given adequate consideration to these factors. However, recent advances in all of these areas, particularly discoveries regarding the mechanisms involved in host searching behavior, have been highly encouraging. The discovery of kairomones and the preliminary development of techniques for their use in managing the field behavior of *Trichogramma* is of particular significance. These developments greatly increase the probability of successful employment of *Trichogramma* in the near future.

Dewinging Beneficial Insects

Entomologist Carlo M. Ignoffo of the Biological Control of Insects Research

Unit released beneficial species of de-winged insects on soybean plants in green-house and field studies. The scientists reasoned that the efficacy of beneficial insects would be increased by partially immobilizing them. They hypothesized that perhaps chemicals inhibiting wing development could be found and wingless beneficial insects could then be mass produced and released to control harmful insects. Induc-

The adult ladybug, unable to fly, still maintains her voracious appetite as she prepares to eat a caterpillar egg.
(Agricultural Research, USDA)

ing genetic mutations of winglessness or selecting mutants that are naturally wingless might even be a more feasible step toward mass production, the researchers suggested. "Our field study with *Podisus maculiventris,* which prey on cabbage looper larvae, showed that 84 percent of the dewinged predators remained at least three days in small soybean plots where they were placed," said Dr. Ignoffo. In other plots where winged *P. maculiventris* were released, only 12 percent remained. In addition, the dewinged beneficial predators laid about nine times more eggs than the winged predators. Eggs from both winged and dewinged eventually hatch into normal wingless young predators which do not develop wings until they become adults, yet prey upon insect pests during their entire lives.

In other soybean plots, the scientists released dewinged *Hippodamia convergens* —commonly known as ladybugs, which feed upon eggs of the soybean podworm. Soybean podworms, also known as corn earworms and cotton bollworms, are serious pests of soybeans. About half of the dewinged ladybugs remained in the plots for at least one day and virtually none of the winged ladybugs remained. Feeding by the dewinged ladybugs reduced the egg population about fourfold.

In a laboratory study, the scientists placed young larvae of soybean loopers on potted soybean plants. Then, as the larvae fed on leaves, they were exposed to groups of dewinged, beneficial, wasp-like parasites *(campoletis flavivinvta).* These parasites killed about 85 percent of the soybean loopers, reported Dr. Ignoffo. In contrast, winged *C. flavivinvta* parasitized about seven percent of the larvae.

IPM Techniques

This section shows the range of methods used in Integrated Pest Management, from plant breeding and interplanting to crop rotation and sex attractants.

Plant Breeding

Eric A. Pillemer and Ward M. Tingey of Cornell University found that certain dry bean plants fend off some pests by arming themselves with sharp, resilient, hooked hair-like structures that entangle, puncture and tear insect enemies. "These microscopic outgrowths, known as *trichomes,* are as effective in limiting insect attacks on beans as thorns and spines are in discouraging larger herbivores from devouring cacti and rose bushes," they explained.

Their study of these plant structures was prompted by an effort to identify unusual physical barriers to insect attack, with the hope that such structures can be bred into valuable crops. "Breeding crop plants for genetic resistance to insect pests is an ecologically compatible alternative to chemical pesticides," they said. The Cornell scientists added that although there is much evidence indicating that plants have a stock of natural chemicals to protect themselves from insect attack, little work has been done to explain plants' physical defenses.

Their work began as a cooperative effort with the International Center for Tropical Agriculture in Columbia to find a control for leafhoppers on beans, an important protein crop in Latin America.

"During studies of feeding damage we observed immature and adult leafhoppers clinging to leaves of certain bean varieties," Pillemer and Tingey said. "Using a dissecting microscope with a magnification of 60 times, we found leafhoppers physically impaled on leaf hairs. As leafhoppers are very active, we concluded that the insects became entangled in the hairs as they moved rapidly over the leaf surface.

"Leafhoppers captured by their leg parts stop feeding and struggle to free themselves from the trichome, frequently becoming impaled on other trichomes during the escape struggle. If unable to escape, the insect usually dies because of dehydration and starvation."

Agricultural scientists at Rutgers University are involved in plant breeding to develop fruits and vegetables best suited to local soils and climates, and best equipped to resist local insects and diseases. Mite-resistant eggplant, the disease-resistant Ramapo tomato, disease-resistant asparagus and improved peach varieties are some of the successful varieties discovered by plant breeders there. They have also found an apple germ plasm which has a high resistance to mildew, fire blight, rust and some minor diseases, and even to mites. According to Dr. Fred Hough of Cook College's Agricultural Experiment Station at Rutgers, "It will take another lifetime, but we are on our way to an apple industry with very limited use of chemicals. It's a long, expensive chore. But one year's commercial orchard spray bills in New Jersey would practically pay for all our work over the past 30 years. So it depends what you consider expensive, doesn't it?"

Breeders and entomologists at the

Texas Agricultural Experiment Station and Agricultural Research Service developed an insect-resistant cotton. When tested in Tampico, Mexico in 1975 under an extremely heavy bollworm-budworm attack without insecticidal control for these two pests, the improved cotton produced 1,-532 pounds to the acre compared to 321 pounds by a commercial cotton, grown as a control. The results were reported by M.J. Lukefahr at the Brownsville (Texas) Agricultural Experiment Station.

Interplanting

Researchers at the Tropical Agricultural Research and Training Center in Turrialba, Costa Rica have discovered that when basic food crops such as corn, beans and cassava are grown together on the same plot, the size of the harvest per unit of land may be increased from 100 to 400 percent. Several years ago, scientists associated with CATIE's Small Farms Project observed that there seemed to be less insect damage when crops were interplanted. This opened up the prospect of controlling insect populations with less use of chemical insecticides. Working under the direction of Small Farms Project horticulturist Dr. Miguel Holle, University of Michigan ecologist Stephen Risch is studying beetle populations in sole-planted and interplanted plots. His research, supported by a Soil and Health Foundation grant, tends to support the theory that interplanting helps keep insect populations under control. Risch is concentrating on four beetle species that are capable of causing catastrophic damage to food plants unless kept in check.

Risch has planted three basic food crops—corn, black beans and bush type winter squash. Some of his 12 plots have sole-planted crops; others have crops interplanted in various combinations. Using stationary nets, Risch captures beetles as they immigrate and emigrate from his plots in search of host plants. But most of his work consists of checking individual plants for beetles and beetle damage, and keeping records of beetle populations and activities in sole-planted and interplanted plots.

Crop Diseases Controlled by Various IPM Methods

University of California (Riverside) scientists have discovered that reflective mulches prevent insect spread of disease. Flying green peach aphids spread watermelon mosaic virus among squash, cantaloupe, watermelon and other plant material by inserting their needle-like mouth parts into the plant, thereby inoculating it with the virus. According to UCR Extension Vegetable Specialist Hunter Johnson, Jr., watermelon mosaic virus is the most serious disease in cucurbit fields of California's Imperial Valley and other desert areas nearby.

Use of aluminum foil and white plastic as mulches to reflect the sky's light confused these aphids in a beneficial way. "The aphids see ultra violet instead of the blue-green light of plants. In effect, they get a signal to 'keep flying' instead of landing. Whatever the exact explanation is, we've found that aluminum foil keeps 96 percent of the aphids out of squashes for a whole season," reasoned Extension Entomologist Nick C. Toscano, Riverside.

The researcher found aluminum foil to be the best preventive of virus spread, reducing virus incidence by more than 90 percent until May and averaging 85 percent reduction for the whole season. However, both of these mulches resulted in a 45 percent increase in yields, compared to the untreated plants, and early in the harvest, yield increases were 86 and 76 percent, due to aluminum and plastic, respectively.

UCR entomologist Jeffrey A. Wyman, head of this research project, concluded, "Such intensification of early production would be doubly beneficial by concentrating peak production when the market is likely to be higher and by allowing early termination of harvest. This in turn would save on labor and reduce late-season virus infection. Most virus spread occurs late in the season. The sooner the grower can cease operations, the better off he is." Insecticide treatments cannot prevent the spread of watermelon mosaic according to the researchers, because the aphid lands and transmits the virus rapidly, and instead of staying in cucurbits, immigrates intermittently from other crops and from weeds.

The discovery of a biological control for Crown Gall by an Australian researcher has been verified by two University of California plant pathologists. Crown Gall is a bacterium that causes an infection which produces rough heavy growths at the base of the trunk, and on the roots, thus clogging up the tissues which carry water and food up and down the trunk. Stunted trees are the result of this disease.

Aluminum foil on zucchini squash provides protection from aphid-treated mosaic viruses. Aphids see blue light of reflected sky and fail to land on the squash.
(California Agriculture)

The Australian researcher, Dr. Allen Kerr, found that dipping seedlings into a suspension of a particular non-virulent strain of *A. tumefaciens,* which itself did not actually cause any disease symptoms, enabled trees to grow healthily, even if the soil in which they grew was infested by the disease-producing bacterium.

Managing sugarbeet yellow and potato leaf roll diseases is the topic of a pilot research project being developed for the Pacific Northwest by scientists at Yakima and Prosser, Washington. The lead scientist is Dr. George Tamaki. These virus diseases are transmitted by the green peach aphid. In turn large populations of viruliferons aphids are generated on weed hosts found primarily on the floors of cultivated orchards and on the banks of drainage ditches. Further, these weeds serve as the reservoir for many virus diseases of vegetable, ornamental and sugarbeet crops. The major element of the visualized pest management system is the permanent replacement of weed hosts of the green peach aphid with grass cover or by cultivation.

Area-wide programs to manage the green peach aphid have already been developed in Washington, Oregon and Idaho, reports Dr. Klassen of the USDA. An important component of these programs is the use of insecticides to suppress the aphids on peaches and other overwintering hosts in spring before they can infest crops. Although the amount of insecticide required is very small, nearly the entire aphid population is treated. Thus, the hazard of developing insecticide resistance seems high. Another component has been the burning of weeds in drainage ditches.

However, this is merely a temporary control and must be repeated each year.

To conduct this pilot research project, a team has been assembled including an orchard weed specialist, an aquatic weed specialist and two research entomologists. The effectiveness of different vegetation management practices in the orchard for the suppression of the green peach aphid are being evaluated. In so-called "clean cultivated orchards," one or two cultivations are performed in May or June rather than just in April and July. Grass cover alone and combinations of grass cover and herbicides are also being evaluated for suppression of aphids in orchards. Similarly various vegetation management systems are being evaluated for use on ditchbanks. Emphasis is placed on a system of converting weedy ditchbanks into low-growing grassed areas.

Soil Fertility and Rotations

Proper soil fertility is advantageous in controlling soybean cyst nematodes. Dr. Woody N. Miley, soils specialist with the Arkansas Cooperative Extension Service, has found that maintaining the soil pH between 6.0 and 7.0 helps to keep phosphorus available, prevents manganese toxicity and maintains a favorable condition for soybean nodule bacteria. A two-year experiment at the University of Missouri examined the effects of nematicides and fertilizer on a soil that tested low in potassium. The study found that applying the nematicide did not increase the yield as much as adding potash. Cyst nematodes were not the major problem. A combination of potash and nematicide produced the highest yield for soybeans susceptible to cyst. But with the cyst-resistant varieties, the nematicide plus potassium did not increase the yield more than potassium alone. Rotating soybean land with cotton and grain crops can also reduce cyst nematode populations. Any rotated crop increasing the soil organic matter is likely to improve the relationships between the soil, plant and moisture. That will result in reduced damage by soybean cyst nematodes. Resistant varieties, nematicides or crop rotation will not increase soybean yield unless proper soil fertility practices are followed on soils infested with cyst nematodes.

Crop Rotation for Managing Nematodes, Diseases and Weeds in Multiple Cropping and Minimum Tillage Systems is a pilot test now underway at Tifton, Georgia. This is a co-operative undertaking between the Georgia Coastal Plains Experiment Station and the Agricultural Research Service. The coordinator of the project is Dr. Clyde C. Dowler. A total of 15 scientists are substantially involved in the project representing the disciplines of weed science, horticulture, nematology, plant pathology, soil science, entomology, engineering, economics and extension.

Although rotation of crops is a traditional method for suppressing pests, the use of sequential cropping with agronomic-horticultural crops to reduce chemical inputs for pest control is a new concept. Some cropping sequences have been shown to be two to four times as effective in suppressing weeds as others. Previously monocropping with peanuts was shown to more effectively suppress nematode populations than do monocrops of corn, cotton or soybeans. However, this is the first experiment which simultaneously measures effects of crop sequences on weeds, nematodes, insects and plant diseases. The two major objectives of this pilot test are (1) to evaluate intensive cropping rotations with new and improved pest management systems for controlling weeds, nematodes, insects and diseases, and (2) to evaluate the effects of various tillage methods on pest management and on crop production. Applications of chemical pesticides are minimal to optimal but not fully intensive because of the study of cropping sequences and tillage practices on pests. Tillage methods and seedbed preparation are being studied within each cropping sequence. Cropping sequences will involve combinations of horticultural and agronomic crops planted in the year rotations as follows:

a. field corn (March 15), sorghum (August 1), watermelon (March 15), lima beans (July 15);

b. turnips (February 15), peanuts (April 20), cucumbers (August 25), sweet corn (March 20), and soybeans (July 15); and

c. tomatoes (March 20), cucumbers (July 1), turnips (September 1), corn (April 1), and southern peas (September 1).

The four tillage systems in each cropping system are: disc harrow only, deep plowing, in-row subsoiling and discing, and in-row subsoiling and planting (minimum tillage).

—Dr. Waldemar Klassen

Sex Attractants and Traps

Laboratory scientists are synthetically duplicating sex attractants, called pheromones, in about 40 individual species of insects. Pheromones are used in the reduction of insect pest populations on crops such as fruit and cotton.

One way in which the sex attractants are used as control methods is by placing a small amount of a particular attractant in a vial that is inside a trap covered with a sticky substance. These traps are placed in orchards, vineyards and in cotton fields.

In "Modern Methods of Pest Control," Roscoe Krauss discusses various ways sex attractants have been implemented as control devices (Volume 10, 1977 of *New York's Food and Life Sciences*):

"Monitoring insect populations by means of attractants placed in traps has been quite successful in commercial orchards. Results have been: fewer pesticides used, satisfactory control and a cleaner environment.

"Equally exciting is the potential use of these attractants to confuse the males so that they are

The small capillary tube hanging from the center of this insect trap releases a controlled amount of a sex attractant blend which permeates the orchard and confuses the males so they cannot find females for mating.
(Cornell University, Agricultural Experiment Station)

unable to find the females for mating. This mass-confusion technique is being studied intensely in New York, and in California it has been used successfully on 25,000 acres of cotton to control the pink bollworm. A promising possibility for affecting several species is to mix a pheromone (attractant) blend. By disrupting the males' mating habits, populations could be brought to levels that remove them as an economic threat.

"Insects have other types of pheromones than those involving sexual attraction. Cornell scientists are studying a warning signal emitted by the secretion of minute portions of a chemical from an aphid attacked by a predator. Aphids are notorious carriers of virus diseases causing thousands of dollars of damage annually to New York crops. Within the past five years, Cornell scientists have identified the alarm pheromone of two sub-families of aphids and, just recently, the alarm chemical used by a third species--the one most damaging to forage crops.

"What actually happens in nature is that when an aphid is attacked, it sends out a "dying gasp" signal to its companions. If the colony is on an apple leaf, the other aphids drop to the ground. Research has shown these alarm pheromones to be of short duration, which means that within seconds the aphids start the long climb back up the host plant to resume feeding, though most of them probably don't survive the trip. However, Cornell scientists have synthesized several simple analogs of these pheromones that are much more stable and may therefore act as a permanent repellent."

Pheromone has also been found to be an effective pest control outside university research centers. A cotton grower in Arizona reported at the 1978 Western Cotton Production Conference that by spraying the sex hormone in his cotton fields, the female pink bollworms became confused, having no mate to encounter when they recognized the sex lure in the air and therefore could not lay viable eggs without mating. This farmer was particularly pleased with his fields which yielded more than two bales of cotton per acre.

Another grower in Thatcher, Arizona described how he spent less than $10.00 per acre for cotton insect control in 1977. The low cost was explained by the early season suppression of pink bollworm, followed by field scouting to keep an accurate check on beneficial and harmful insect numbers, help from predators and parasites and use of insecticides solely as a last resort. This grower reported, "On my 715 acres of cotton, I invested $4.50 per acre in early season pink bollworm suppression using paper cup traps baited with pheromone. Then, I spent another $2.00 per acre to have my fields scouted on a weekly basis. Finally, application and my insecticide use averaged $2.80 per acre. So, I come up with a total figure of $9.30 per acre." This cotton grower is a board member of the Graham County Pink Bollworm Committee which represents 80 farmers who have teamed up since 1969 to operate a pest management program. For the past three years, the committee has cooperated with Dr. Roger Huber, University of Arizona entomologist, in early season pink bollworm suppression attempts. It involved placement of carry-out coffee cup traps mounted on a stick in 15,500 acres of cotton, during the 1977 growing season. Spotted at an average 4.5 traps per acre, each contained a synthetic sex attractant to lure male pink bollworm moths. In the cup bottom, a dash of cotton picker oil snared and drowned moths which blundered into the trap seeking a mate.

Sex attractants are used effectively to suppress screwworm fly populations. Scientists at the Screwworm Research Laboratory, Mission, Texas use a two-by-three-inch paper cylinder trap, containing an attractant, bait and minute amount of a nonpersistent insecticide. The attractant simulates a festering flesh wound. Cylinders are dropped from aircraft at the rate of about 20 per square mile.

The major role of the "screwworm fly killer" is to reduce the native population to levels where artificially-sterilized males from mass rearing facilities in Mission, Texas will be more successful in eradicating natural populations.

A pilot test, led by R. W. Miller of the Agricultural Research Service in Beltsville, Maryland, uses sticky panel traps instead of chemicals to control face flies, a major pest to beef and dairy cattle. Female face flies feed on secretions around the eyes and nostrils. Two habits of the face fly offer opportunities for population management.

The first is that the eggs are laid exclusively in cattle manure. The other is that male and female adults are attracted to white plywood panels in large numbers. The panels are coated with "stickem" or with an insecticide.

These traps are used to catch face flies, a major pest of cattle, near feeding and watering grounds. Attached to steel fence posts three feet off the ground, the height at which most face flies are found, these traps have reduced face fly numbers by as much as 70 percent in experimental pastures at the Beltsville Agricultural Research Center in Maryland.
(Agricultural Research, USDA)

New Directions in Weed Control

Although North American farmers apply about $1 billion worth of herbicides on 200 million acres annually, a great many weed specialists stress the tremendous weaknesses in current control strategy. Ellery L. Knake of the University of Illinois, for example, warns about great problems if "we confine our thinking to a single control method," i.e., herbicides.

Biological weed control is the deliberate use of natural enemies (plant-feeding and disease-causing organisms) to reduce the densities of weeds to economically or aesthetically tolerable levels. "This weed control method," wrote Richard D. Goeden of the University of California, Riverside in *Weeds Today,* "has been successfully employed against a limited number of weeds on a large scale for many decades."

The USDA's western region Agricul-

California spends more than $500,000 per year to impede Russian-thistle invasion which nevertheless persists. The Pakistani moth, C. parthenica, shown here, lays its eggs on Russian-thistle stems and leaves. The hatching larvae feed into the leaf and interfere with the growth of this weed.
(Agricultural Research, USDA)

tural Research Service in Albany, California has an extensive program. Lloyd A. Andres, Location Leader in Albany, lists these ongoing biological control projects for the following weeds:

Leafy spurge	—*Euphorbia esula*
Musk thistle	—*Carduus nutans*
Russian thistle	—*Salsola iberica*
Halogeton	—*Halogeton glomeratus*
Tansy ragwort	—*Senecio jacobaeae*
Puncturevine	—*Tribulus terrestris*
Rush skeletonweed	—*Chondrilla juncea*
Mediterranean sage	—*Salvia aethiopis*
Diffuse knapweed	—*Centaurea diffusa*
Spotted knapweed	—*C. maculosa*
Yellow starthistle	—*C. solstitialis*
Italian thistle	—*Carduus pycnocephalus*
Milk thistle	—*Silybum marianum*

According to Professor Goeden in his *Weeds Today* paper, insects have had the greatest use as natural enemies in biological weed control, with records of such going back more than a century. "Some results with biological control have been spectacular," he noted. "The two most frequently cited examples are the destruction of prickly pear cacti on millions of acres of rangeland in Australia by an imported moth, and biological control of St. Johnswort on about two million acres of rangeland in California by an imported beetle. These examples also illustrate the advantage of biological control methods essentially providing a permanent solution to a weed problem. . . . Most biological control programs involve weeds of foreign origin that were either accidentally or intentionally imported. . . . Most weeds targeted for biological control are terrestrial, perennial, alien species infesting relatively undisturbed, extensive land areas (range-land, pastures, forests, roadsides and recreation lands) of low unit value that makes widespread application of chemical or mechanical control uneconomical. . . . The importance of biological weed control is likely to increase with the maturation of weed science."

University of Connecticut's R. A. Peters is examining acceptable economic levels of weeds—reducing herbicide rates and seeking early control of crabgrass. After the corn shades the ground, late-germinating grasses have little effect on yields. The thin cover of dead grass, Peters found, keeps the soil in place.

In a report on future weed control options, Rex Gogerty wrote the following in *The Furrow:*

"Some plants can compete by 'poisoning' neighboring plants. Through a mechanism called allelopathy, their plant roots secrete substances into surrounding soil and inhibit growth of nearby plants. Allelopathy is known to exist to some degree in such commonly grown crops as corn, soybeans and cereal grains. Even as the plants decay, their residues may continue to exude toxins. Researchers Alan Putnam of Michigan State University and William Duke of Cornell have found one cucumber variety that inhibits 75 percent of normal nearby weed growth. Scientists theorize that it might be possible to increase allelopathy through plant breeding."

Other researchers at the University of Arkansas are getting encouraging results by inoculating a fungus on curly indigo (also called northern jointvetch), a pest that clogs drainage ditches and rice paddies in the South. The fungus kills the weed without harming other beneficial plants. In his report in *The Furrow*, Gogerty predicts that pathogenic weed con-

trol "promises to become much more useful in the future. Scientists hope to use it, for example, against spurred anoda, a troublesome weed in southern cotton and soybean fields. In tests, adding two naturally occurring fungus diseases to fields reduced populations of spurred anoda to zero."

Gogerty also mentioned some other new weed control options: weed-eating fish (grass carp) and mammals (manatee) and insects *(Bactra verunta);* mechanical and electronic aids—"The Zapper," a tank-size weed shriviler from a Texas firm, and Lightning III, a tractor-drawn electronic weed-shocker from Mississippi; and light sources that affect light-sensitive weeds.

Several firms now make underwater weed cutters and pond sweepers. In Bradenton, Florida, the Kee Manufacturing Company's cutter is comprised of two serrated blades, while Joe Morreale in Rich-

Heavy mats of alligatorweed in the bayou country challenge these fishing camps. Many southern waterways are being cleared by the tiny South American flea beetle, *Agasicles hydrophilia,* whose natural host is alligatorweed.
(Agricultural Research, USDA)

mond, Illinois offers building plans for a "pond sweeper" for gathering weeds as they are cut.

Richard Thompson, who organically farms 300 acres in Boone, Iowa, explains how he controls weeds without relying upon herbicides:

"I feel that the most important item in weed control is crop rotation. The three years of row crop is followed by oats and then hay. During the two years that the field is seeded down to legumes and some grasses, no weeds should be allowed to go to seed. This is not hard to accomplish. The oat field will be combined for oats in July and then clipped in September before any weed seeds are mature enough to germinate . . . The hay field the following year will be cut three times with no weed seed developing. This kind of management is real hard on the grassy type weeds . . . I feel our giant foxtail problem began when 2,4-D was introduced and used to kill the horseweed, hemp and other tall weeds in the fence rows around the fields. This allowed the foxtail in the fencerow to grow. Its competition had been killed by the 2,4-D. I can remember how great it was to ride the tractor around the edge of the field and hold the boom nozzle to spray the weeds. Within several hours, the broadleaf weeds would start to wilt. However, a greater weed problem was created. The broadleaf weeds that moved out into the field were not difficult to control with mechanical cultivation. This was not true with giant foxtail as it moved in from all sides of the field. About this same time, oats and hay were being eliminated from most rotations, which accelerated the giant foxtail grass problem. This same situation is happening again with the latest herbicides. Certain weeds are resistant or are developing resistance and these weeds like milk weed, ground cherry and others are increasing in numbers.

"The way we handle weeds in the fence row is with a fence row mower. All the weeds (grasses and broadleaf) are mowed off once during July.

The timing is important so that the weed seeds are not mature enough to germinate, but the plant is far enough along in maturity that it will not try to produce another seed. Timing and types of cultivation are important tools in weed control. For corn and soybean production, the soil is disked as early as possible in the spring to get weeds started to grow. In our situation, this would be sometime in early April. A second disking would take place two or four weeks later, depending on weather conditions. Just prior to planting, a field cultivator is used at a 4- to 6-inch depth, using 9-inch sweeps on 6-inch spacing. This operation will leave the top surface with more clods which is a poor weed germinating condition. This semi-deep field cultivation does cut off deep rooted weeds like milk weeds, pond weeds, volunteer corn, cocklebur and sunflowers. Since this is the last tillage operation and is late in the season, it has been very helpful in weed control. With weather permitting, all pre-planting tillage and planting operations are attempted at the proper stages of the moon.

"After the crop is planted, the field is harrowed just prior to the seed emerging from the soil. This harrowing drags out the small white seedlings. One week later, the field is gone over with a rotary hoe. The second rotary hoeing should be done seven days later, All harrow and rotary hoe operations should be done before *any* emergence of weed seedlings. If the field is starting to look green between the rows, it's too late for these two tillage methods.

"The first cultivation usually comes one month after planting. The second cultivation may be the final cultivation if the weather has cooperated with the timing schedule. If the weeds have gotten away from us in the corn field, the third cultivation is done with disc-hillers on the cultivator to throw as much dirt as possible to cover the weeds in the row. This extra soil around the roots causes an extra set of brace roots to develop. Since we have ridged the corn plant, lodging has been almost zero.

"The number of trips over the field prior to planting are about the same as my neighbors. They will disk once to smooth the field, the next disking will come when they put on their herbicide and then a spring tooth operation ahead of planting. After planting they will rotary hoe once and cultivate once. Their herbicides will save the third or fourth trips, one with the harrow, one rotary hoeing and one or two cultivations.

"Our planting dates are later than conventional farmers in the neighborhood. However, these planting dates are like they were before herbicides became available. Oats planting time for us is April 1st to 7th, corn is May 5th to 15th, beans is May 20th to 30th. This balanced work load is sure better than trying to plant the whole farm to corn by the last of April."

Strawberry producers in Nova Scotia have found geese to be effective weed controllers in their fields at a saving of $30.00 to $100 per acre. Ralph McQueen, an animal nutritionist at the Canadian Agricultural Research Station at Fredericton, New

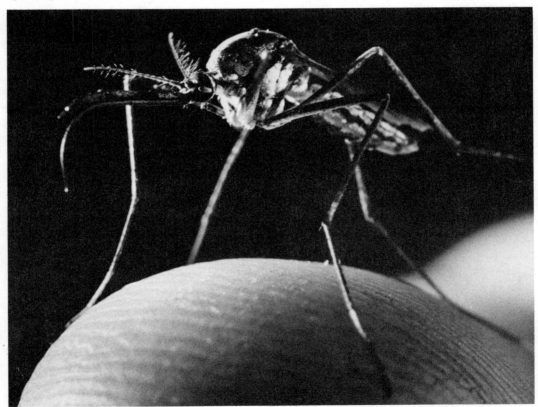

Researchers are studying the potential of this biological control agent, a large non-biting mosquito, Toxorhynchites rutilus rutilus, for the control of the yellow fever mosquito. Disease-bearing mosquitoes have developed resistance to extensively used pesticides.
(Robert Barclay, USDA)

Brunswick, believes that geese have a modern potential for farmers. They grow faster than any other domestic fowl and are more resistant to disease, need only shelter from wind during winter in most Canada locations and do well in pastures.

A system for *Total Weed Population Management for Irrigated Farms* is being developed by Colorado State University, led by Dr. E.E. Schweizer at Fort Collins. Weeds are a major pest of irrigated crops, in part, because irrigation waters facilitate the dissemination of seeds and other propagules, so a system is needed to reduce the tremendous weed seed reserves. For the seven principal genera of weeds in irrigated farms in Colorado, the number of weed seeds per acre ranges from 1,500,000 to 3,200,000.

The total population management system consists of (a) use of weed-free crop seed, (b) use of screens to remove seeds and other propagules, from irrigation water, (c) control of all weeds on ditchbanks by flaming or cultivation and (d) the use of herbicides on an "as needed" basis. In the standard system, herbicides are used routinely along with cultivation to the extent that it can be accomplished by the farm manager.

Several pilot research projects, supported by the Agricultural Research Service, deal directly with weed control on rangeland in the West and Southwest. According to Dr. Klassen, only two percent of federally-owned range is in excellent condition and only 15 percent is in good condition, while 33 percent is in bad condition. "Forage production on western rangelands must be increased by 96 percent by the year 2000 in order to assure level beef consumption in the United States," he notes. The integrated program of range improvement and grazing management emphasizes replacement of weedy species with desirable forage plants. This investigation is being conducted on the Gund Ranch, a working cattle ranch owned by the University of Nevada, Reno.

Where Do We Go from Here?

For Farmers

"We are on the verge of transforming insect control from a system of science and half guesses to one based primarily on facts, in which the promotion of insecticides will no longer be a decisive determinant of what is to be done."

Carl B. Huffaker
University of California

"Petroleum-based pesticides have become, and will continue to be, dramatically more expensive. Eighty percent of the billion pounds of pesticides used in the United States each year are petrochemically based—that is, the active ingredient is a petroleum derivative. This figure does not include pesticides whose production or extraction processes require petroleum-based solvents, nor does it account for the use of petrochemicals as 'inert' ingredients in non-petrochemical pesticides."

Steven D. Jelinek
Assistant Administrator for Toxic Substances
Environmental Protection Agency

An intensive effort needs to be made to break through the communications barrier which currently blocks the flow of pest management information to the farmer. Too many times, a stop-gap pesticide application is chosen when pre-planning for integrated controls would have been preferable.

To pave the way for a "system based on facts," a series of *Demonstration Days* should be sponsored by the Extension Service on farms where Integrated Pest Management methods have been effectively used. Farmers learn from their peers, and —for example—New England apple growers would clearly benefit from a visit next fall to the 110-acre orchard of Ed Roberts. His use of monitoring traps last year saved him the expense of two sprayings, "and that's money in the bank."

Another way to help get the word out to farmers would be through short courses and workshops on how to use IPM on the farm. The meetings should be regionalized as much as possible, so speakers—IPM consultants, extension entomologists and researchers—can specifically discuss pest problems facing those farmers in the audience. Around the country, there are people like Ray Frisbie in Texas, Don Mock in Kansas, Ron Prokopy in Massachusetts, Jim Tette in New York, George Bird in Michigan, to name a few—persons whose knowledge would undoubtedly be most valuable to thousands of growers.

Many projects have indicated the special value of IPM concepts to small farmers. In New York State, Jim Tette, project manager for the IPM program for the agricultural experiment station there, noted how his state's small farmers benefited from the program. Dr. Tette reports a continuing "successful spread of IPM principles and practices."

For the first time in 1977, New York implemented a training program to directly provide growers with a way to conduct IPM on their own farms. Dr. Tette also reported another first for his state's IPM project: "In-depth IPM workshops were made available on a trial basis to fieldmen, agents, growers and new IPM personnel. Those sessions included information on pest identification, application techniques, and pest control strategies. This was followed by weekly IPM bulletins and in-season training sessions. The success of this trial training program has spawned a series of IPM workshops across the state scheduled for this winter."

In many states, more grower cooperatives are establishing IPM services for their member farmers. This development will also eliminate more and more pesticide applications.

For the New IPM Industry

"A corps of highly trained professionals will be needed to monitor the major features required. A weather network designed and computerized to satisfy the needs for modeling events throughout the nation is needed. We have seen how such a network is effectively used in insect pest management in Michigan. We have seen how a telecommunication network, tied into a data bank of pest incidence, crop conditions and pest control tactics can be used to update our traditional extension service. . . .

"The chemical industry has for too long dominated the pest control scene, and this has resulted in an almost complete departure from some of the older more ecologically based methods of pest control. A virtual army of pesticide salesmen have in some parts of the country practically replaced the traditional dependence of the farmer on his university for advice.

INTEGRATED PEST MANAGEMENT

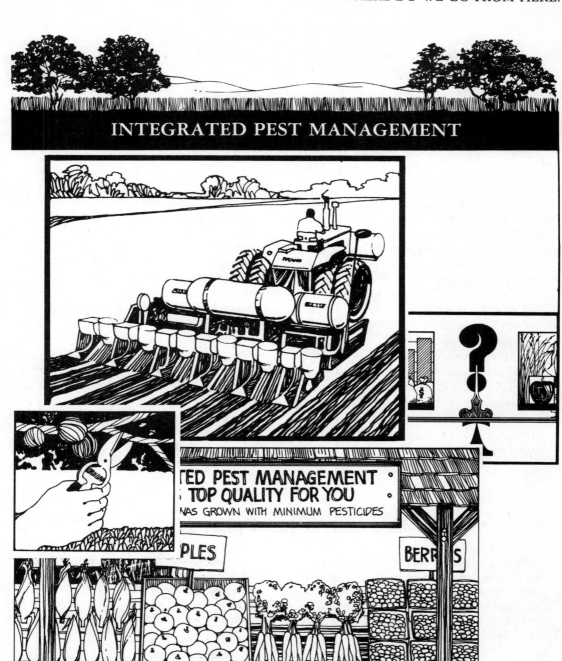

There must be some way that this can be corrected. We should put a force of independent professional biologist-agriculturists in the field to do the necessary monitoring and assessment of the need for treatment and to ascertain what measures, if any, are best."

Carl B. Huffaker
University of California

The "corps of highly trained professionals" is emerging, but overcoming the status quo is a fierce task. The long-term interests of farmers, consumers, policy-makers and environment would be served if the economics work out for various segments of the IPM industry.

Presently, the support system for that industry consists of researchers, scouts, propagators of beneficial insects, computer technology specialists, paraprofessionals and professionals along a continuum that stresses management instead of pesticide application.

The more visible members of the IPM industry become, the better their chances for economic success. One way to gain such visibility is through participation in short courses and workshops, as well as Demonstration Days, sponsored by the Extension Service, farmer cooperatives and others. Leading consultants in the IPM field could be selected to discuss experiences of their client farmers, as well as the ways IPM can be implemented.

As organizations such as the Applied Insect Ecologists and the Southern Alliance of Agricultural Consultants Association and others grow in size, there will be a number of positive effects. One will be a greater awareness by deans of our Agricultural Colleges that career opportunities exist for students. Internship possibilities will become more widespread—not only for entomology students, but for soil science, agricultural engineering and majors in other areas as well. An interdisciplinary approach to IPM is critical, and will become more and more obvious as jobs open up for students with intimate knowledge of how cropping systems and soil fertility relate to pest management.

For Consumers and Policy-Makers

"We need to develop and use alternative tactics in Integrated Pest Management systems. We need to make sure that pesticides used in our programs meet the criteria of appropriateness and safety. And we need to constantly keep in mind that we serve *all* segments of the public —gardeners, small farmers, commercial farmers, forestry, households, food and fiber handling, storage and marketing enterprises. . . . The full support of our research effort will be behind biological controls."

Rupert M. Cutler
Assistant Secretary
Conservation, Research and Education
United States Department of Agriculture

"It is the policy of the United States Department of Agriculture to develop, practice and encourage the use of Integrated Pest Management methods, systems and strategies that are practical, effective and energy-efficient."

> Bob Bergland
> Secretary
> United States Department of Agriculture

"I am instructing the Council on Environmental Quality, at the conclusion of its ongoing review of Integrated Pest Management in the United States, to recommend actions which the federal government can take to encourage the development and application of pest management techniques which emphasize the use of natural biological controls like predators, pest-specific diseases, pest-resistant plant varieties and hormones, relying on chemical agents only as needed."

> President Jimmy Carter

"Despite affirmations of support from high officials, the federal government's program in biological control—which points the way out of the mess created by indiscriminate use of chemical pesticides—remains in disarray. . . . The verbal commitment to biological control has already been made by Secretary Bob Bergland and some of his chief assistants. Nationally, it is a popular concept. But the nagging feeling persists that USDA may be too disorganized to carry out its mandate."

> Frank Graham, Jr.
> Audubon Magazine

The absence of a profit incentive is perhaps the *major* obstacle to the future of IPM. While the fragmentation of government agencies and universities continue to frustrate policy-makers and must be addressed, we believe that even that formidable problem could be overcome as IPM economics improve.

Heretofore, the incentive to use IPM has depended strictly on whether the farmer would reduce pesticide costs sufficiently to warrant the services of an IPM consultant. No attempt has been made by the grower to use IPM as a *marketing tool.* Yet the marketplace is the one area where the rapidly growing number of consumers can show their support for reduced pesticide use. How would such consumers react if offered the opportunity to buy produce grown under IPM conditions? "The chances are many would seek out such produce and might even be willing to pay a premium price for it," forecasts direct marketing expert Don Cunnion. In developing his reasoning in Chapter 7, The Consumer and Pest Management Decisions, Cunnion writes:

"IPM has its greatest marketing advantage in the areas of fruits and vegetables sold for the fresh market. That's because of the growing awareness among consumers of the hazards of chemical pesticides. More and more, consumers are voicing their concern that 'too many chemicals are being used to produce our food.'

". . . A highly visible sign above the displayed food, for example, could explain what IPM is all about: 'IPM stands for Integrated Pest Management. This means fruit grown with a minimum use of pesticides—only enough to assure top quality for you'."

It seems clear to a great many concerned people on farms and in cities and suburbs that the support for IPM is overwhelming. What remains obscure, however, is how to translate that support into dollars and cents for the farmer or other pest management decision-makers and for the persons who can make IPM work. The American consumer and voter must and can provide the missing elements.

Federal as well as state governments could stimulate the market demand for IPM services by making greater use of IPM. Opportunities exist for IPM use on forest lands, parks, along highways, on military bases, etc. The government has precedents for using purchasing specifications to stimulate environmentally-sound developments. In the early 1970's, "specs" called for recycled fibers in paper supplies to increase the demand for the recovery of waste paper. More recently, Congress has proposed legislation to require solar energy devices in our embassies overseas. By specifying IPM measures, the government could help to provide a broader market base for companies in the IPM field.

For IPM Researchers

"Looking back, it was very good that we started as we did with a 'think tank' situation in which we could develop new concepts and plan specific projects. We had time to hammer out a sturdy theoretical framework. This enabled us to plan a general strategy for a set of coordinated, well-defined subprojects and to move rapidly to investigations of the structure and functions of our societal institutions. That beginning was also of special importance because we were able to develop a number of creative, cross-disciplinary relationships. These have proved to be a sound foundation for a lasting program."

Herman E. Koenig
Ecosystems Design and Management Program
Michigan State University

"IPM represents a new conceptual approach that sets crop protection in a new context within a crop production system. Many components of IPM were developed long ago, but IPM as now conceived is unique: based on ecological principles, it integrates multi-disciplinary methodologies in developing agroecosystem management strategies that are practical, effective, economical and protective of both public health and the environment. The early efforts of crop protectionists to control pests with ecologically-based cultural methods were not satisfactory; consequently, entomologists, plant pathologists, and later, weed scientists were preoccupied with the discovery of pesticides that were often used not to supplement cultural methods but to supplant them."

Ray F. Smith
Professor of Entomology
University of California, Berkeley

The potential of the IPM system is only beginning to be realized in the eyes of many researchers. Dean L. Haynes of Michigan State University's Entomology Department believes that pest management will not reach full effectiveness until it is possible to map crop insect ecosystems for a broad area and take measures that will provide maximum benefits with minimal use of sprays. Looking ahead further, Haynes foresees other applications for climatic and biological monitoring tied to computer analysis, similar in ways to early warning systems for avoiding forest fires and peak periods of air and water pollution.

The complexities of IPM will require consistent and increased funding for many projects now underway. These projects, such as the ones described in earlier chapters, are clearly opening up new horizons in vital areas of public interest.

We also believe that more research should be done on farms, despite the lack of facilities for replicating plots. By taking the research team to working farms, entire systems could better be judged—including the motivations and procedures of the farm family.

The research work on organic farms done by the Center for the Biology of Natural Systems at Washington University is a fine example of the value of such on-farm analysis. Significant insight into the reasons why farmers decide to use, or not to use, pesticides is gained from the Center's report on "Motivations and Practices of Organic Farmers," as well as Dr. Rosmarie von Rumker's study on "Farmers' Pesticide Use Decisions."

There will be many benefits derived from focusing research attention on the interrelationship of pest management to the total farm management philosophy of organic farmers specifically, and small farmers generally. Such studies into the pest management practices of farmers who are not relying on chemicals should not and need not be carried on as if they were part of a war between organic fanatics on the one hand and rabid pesticiders on the other. The simple fact is that the methods used by organic farmers and their results constitute *valid models* for careful observation as we seek to develop better pest management systems for the future.

In this book, we have made a determined effort to explain why more Americans must be involved with pest management decisions. Many more persons may soon be working for firms which provide farmers with IPM services. Many consumers can soon be using more of their purchasing power for foods grown with IPM methods.

The time has come to use the mounting knowledge to reduce pesticide use in the United States, and throughout the world. It is more than time to witness a downward trend in pesticide sales. There is no excuse to delay any longer.

Biological Pest Control Agents

"Great fleas have little fleas upon their backs to bite 'em,
And little fleas have lesser fleas, and so *ad infinitum.*

And the great fleas themselves, in turn, have greater fleas to go on;
While these again have greater still, and greater still, and so on."

—Augustus De Morgan,
A Budget of Paradoxes.

Biological control agents, like pests, occur in a great variety of forms. Those considered in this report are grouped in six general categories: Vertebrates, arthropods (spiders, mites and insects), nematodes, snails and other invertebrates, microorganisms (pathogens that affect invertebrates, plant pathogens and microbial antagonists) and higher plants.

Vertebrates

Vertebrates, both carnivores and herbivores, play significant roles in the control of pest populations. They are important largely in stabilizing population levels

This material is excerpted from the publication, Biological Agents for Pest Control: Status and Prospects. *This is a report of a special study team coordinated by the Office of Environmental Quality Activities. It is published by the United States Department of Agriculture in cooperation with the Land-Grant Universities, State Departments of Agriculture, and the Agricultural Research Institute.*

Dr. Rupert Cutler, Assistant Secretary for Conservation, Research and Education for the USDA, has stressed how the

"USDA recognizes the importance of biological control as an essential part of pest management systems." The report from which the following material is excerpted, Dr. Cutler says, "highlights a number of opportunities for the expanded use of beneficial arthropods, nematodes, snails, microorganisms and vertebrates for suppression of many kinds of pests. The Department will consider how to best implement the recommendations of this report to increase activities among all concerned in the research, development and use of biological agents in the management of pest problems."

rather than in lowering them to nonpestiferous densities. Unfortunately, quantitative data adequate to assess the importance of most species are usually not available. Furthermore, because most vertebrates, particularly birds and mammals, have behavioral patterns far more complex and adaptable than those of invertebrates, they tend to be opportunistic feeders and may neglect pest populations if other acceptable food sources are readily available.

Although insectivorous birds that inhabit suburban communities, diversified farming regions and forests contribute to pest control, their actual impact on the ecosystem is difficult to determine. Similarly, the value of flocks of gulls that prey upon white grubs during spring plowing of farmlands or feed on crickets and grasshoppers during outbreaks should not be discounted. But, in the light of past experiences, introduction of birds for purposes of controlling pests seems inadvisable. The Indian mynah was successfully introduced into Mauritius to control a locust; but the same species introduced into the Hawaiian Islands and Australia has become a pest, and its value in those regions is equivocal. The starling appears to be a useful bird in regions where its population remains low, but in both Great Britain and the United States it is definitely a pest. Displacement of native bird populations, flocking habits and urban roosting behavior make such species locally objectionable even though they may exercise a degree of pest control in certain situations.

Management of potentially useful bird species consists primarily in providing suitable nesting sites and cover to encourage their presence in areas where and when pests occur. However, the territorial behavior of most bird species limits the resident population that can be attracted by such actions during the breeding season, when food consumption is high and pests most abundant. Domestic ducks and geese are sometimes used for weed control in many situations and the former are useful in keeping down populations of noxious snails.

Among the mammals, bats, mice and shrews are probably the most effective insectivores. As with birds, there is little solid evidence of the impact of bats on pest populations. In areas where bats are abundant, they consume enormous quantities of insects, judging from the abundance and contents of feces in their roosting places. Shrews and mice appear to have considerable value as biocontrol agents where they feed on pests in forest habitats. One species of shrew (Sorex cinereus) was introduced into Newfoundland from the adjacent continental area and was successful in depressing populations of the larch sawfly in a climax coniferous forest environment. The mouse (Peromyscus leucopis) has been shown to be an important predator of the gypsy moth in New England. Another mammal actively studied for weed control in the past is the manatee.

Reptiles and amphibians appear to have limited utility in pest control. Lizards account for a considerable mortality of insects, some of which are pests, but their effects on pest population dynamics are not documented. The giant American toad (Bufo marinus) and the European toad (B. bufo) appear to be effective in certain habitats. The former has been shown to be useful against coleopterous pests in sugarcane fields, and the latter is used to control pests in greenhouses.

Of all the vertebrates, fish are undoubtedly the most useful for the control of pest insects, particularly mosquitoes and midges. Numerous species are known to be insectivorous, but the management of fish for mosquito and midge control has been limited primarily to the mosquito fish and the guppy. The use of these two species, and a few others, for control of mosquito breeding in shallow waters and artificial containers is widespread in warmer parts of the world. Unfortunately, the mosquito fish shows some tendency to be selective in its feeding, and may not always be effective in suppressing malaria vector populations if larvae of preferred species are readily available. Neither fish is suitable for human food.

In some situations herbivorous fish such as *Tillapia, Ctenopharyngodon* and *Cyprinus* species are useful in aquatic weed control in rice paddies and artificial fish ponds. Fish of these genera, carp for example, provide food for man; but efficient use of them in a dual role requires considerable management knowledge and skill. Indiscriminate introduction of presumably useful species of fish should be avoided, as with birds, and all introductions should be preceded by critical ecological studies.

In summary, relatively few vertebrates are amenable to management for control of pest populations, although many species, through their natural behavior, undoubtedly play significant roles in keeping pests at low levels.

Arthropods

As a group the arthropods unquestionably predominate both in the numbers of kinds that are pests, and the numbers that play important roles in regulating populations of other animals. The importance of insects as biological control agents is particularly noteworthy, as evidenced by the numerous treatises that deal with their use in biological control programs. In fact, development of principles and practices in the field of biological control derives in large part from studies of insects as natural enemies of insects. Except for insects and mites, the roles of other arthropods as regulators of pest populations are little known. Within the limits of their rather specialized environments, centipedes and scorpions and their relatives play an important part in maintaining a balance of animal life, but as a group are more likely to be regarded as pests of humans than as benefactors. Only the common house centipede, which feeds upon small animals that share its habitat, has any significant reputation as a pest control agent.

Spiders and Mites

Spiders, all of which are predators, occupy an equivocal position. Most are opportunistic feeders, but a few are selective feeders. Some may be classed as pests because of their destruction of beneficial insects; others, particularly those that inhabit vegetation and feed upon herbivorous insects, are beneficial, although the benefits are more as general population regulators than as enemies of pest species. In Japan, spiders are thought to be important in early season control of leafhoppers in rice fields. However, at present, spiders are not amenable to "management" for use as biotic agents because mass rearing techniques have not been developed. The best tactics for maximum exploitation of them would be maintenance or creation of environments (absence of toxic chemicals) that permit their survival.

Mites are important as regulators of pest populations. Predatory species of *Typhlodromus* and *Phytoseiulus* have been shown to be effective in regulating populations of several species of spider mites that are serious pests of cultivated crop plants. Because these predatory mites are not so vagile as the spider mites, which disperse by "ballooning," they cannot regularly be depended upon to regulate populations of their prey; augmentation or reintroductions in pest-occupied habitats may be required. However, inasmuch as spider mite outbreaks appear to be triggered by injudicious use of chemical pesticides that eliminate the predatory mites, a reasonable level of natural spider mite control may be possible in the absence of pesticidal applications or by selective use of pesticides.

Numerous trombidiform mites are predators of insects and other mites, for example, *Allothrombium fuliginosum,* a predator of eggs of the European corn borer, *Cheyletus* spp. that prey on tyroglyphid mite pests of stored products, and *Macrocheles muscaedomestica* that preys on eggs and first instar larvae of the housefly. The last mentioned species seems to have considerable potential for control of fly breeding if suitable environmental management practices are followed with the breeding medium, animal dung.

Although serious work directed toward "management" techniques for mites as biotic agents for pest control has only recently begun, the outlook is promising. Predatory mites of some kinds disperse by attachment to the mobile stage of their host (phoresy), but in general it would seem that their value could be increased by distribution as needed. A red spider mite, *Tetranychus desertorum,* probably contributed to some extent to the biological control of cactus in Australia; the possibility of using phytophagous mites for the control of weeds is being critically investigated.

Insects

Insects may be used as biotic agents for pest control in three basically different ways: As parasites or predators that attack their animal hosts (the pest species), as herbivores that feed on noxious plants, or as genetically modified populations that affect the reproductive potential of "wild" populations of the same species. Although some common features apply to all three methods, it is convenient to consider them separately. Relatively few species have been shown to be amenable to genetic modifications that make them effective control agents, but the number of potentially useful parasites, predators and herbivores is legion. For example, in a single hymenopterous family Ichneumonidae, all members of which develop as parasites of other insects, it is estimated that there are 60,000 species in the world, yet only about 16,000 of these have been named. Our knowledge is equally fragmentary of the world's fauna in other groups of prime importance as biotic agents for pest control. C.P. Clausen's monumental treatise, *Entomophagous Insects,* includes discussions of representatives of 224 families in 15 (or 16, depending upon the classification followed) orders that have adopted, in some measure, the entomophagous habit. Insects are also useful for the control of invertebrate pests other than arthropods, particularly snails.

Entomophagous and Phytophagous Insects. The role of many parasitic and predatory insects is in maintaining a balance among the organisms within the environment they occupy. This is a most important role because it prevents those of their hosts that have the potential to become pests from doing so. Many others, even though voracious feeders, may have little effect on population levels of their prey because they are too opportunistic in their feeding habits or too few in numbers. Dragonflies, that feed on mosquitoes and midges and mantids and scorpionflies, that feed on many kinds of small insects, are in this category. Although a few entomophagous species in other insect orders are useful as pest control agents, this report considers primarily those belonging to the Hemiptera (true bugs), Coleoptera (beetles), Neuroptera (lacewings and others), Diptera (two-winged flies) and Hymenoptera (ants, parasitic wasps, and others).

Insects are the largest group of natural enemies of weedy plants, and host-specific insects have been used in the biological control of weeds for over 100 years. Their value has been demonstrated in many countries, and insects are now under study in over 70 weed control projects in North America alone. They have proven of greatest value against introduced weeds. The weediness of some introduced plants may be partially attributed to the absence of natural enemies in their new habitat. The search for, testing, and introduction of exotic biotic agents for control of these weeds is often an attempt to correct this imbalance. Weed-feeding insects have been used primarily to control perennial or biennial species growing in areas of low disturbance such as rangeland, forest and pasture. Unfortunately, a high degree of disturbance (for example, cultivation) frequently interrupts the insect developmental cycle, precluding their increase to control levels in cropped or similarly disturbed areas. Most of the phytophagous insects thus far used for weed control belong to the orders Hemiptera, Lepidoptera, Coleoptera and Diptera; occasionally insects of the orders Thysanoptera, Orthoptera and Hymenoptera are used.

Grasshoppers are usually considered to be pests, but some are specific enough in their habits to warrant use as agents for weed control. One leaf-feeding species from South America, *Paulinia acuminata,* has been introduced into Africa to help combat the floating aquatic fern, *Salvinia molesta,* in Lake Kariba. *Salvinia* is a serious aquatic weed in parts of Asia and Africa, where it is naturalized from the neotropics.

The Neuroptera include some 19 families, all members of which are predaceous, at least in the larval stages. Some members of two families, the green lacewings (Chrysopidae) and the brown lacewings (Hemerobiidae) are important biological control agents, although, thus far, only some of the green lacewings have been widely used in the sense of being manipulated. The larvae of green lacewings are predators of aphids, small lepidopterous larvae, other small soft-bodied insects and spider mites. The adults of some species are also predators.

Those species of green lacewings that feed on honeydew as adults can be manipulated to concentrate populations in specific areas by using artificial "honeydew" consisting of mixtures of yeast hydrolysates, sugar, and water. Some species of *Chrysopa* are being mass produced in the laboratory and are available for sale as agents for pest control.

The brown lacewings are predaceous on aphids, mealybugs, mites, and various kinds of insect eggs. Thus far little use has been made of species in this family because of the lack of information about their biologies and ecological requirements. However, recent studies indicate that brown lacewings can develop at very low temperatures (0.4–4.1° C), suggesting that they may be useful for very early season control of aphids when other kinds of biological control agents are not yet active.

The importance of Lepidoptera as agents for pest control results from their primarily phytophagous habits in the larval stage. Larvae of Lepidoptera are the most important insects that feed on the dwarf mistletoes, at times causing severe damage to external parts of the plants, and the use of insects offers considerable promise for control of these important plant parasites. A moth, *Cactoblastis cactorum,* with an associated bacterium that invades plant tissue following feeding of the moth larvae, is largely credited with the spectacular control of cactus (*Opuntia* spp.) in Australia. Larvae of the cinnabar moth, *Tyria jacobaeae,* are the most important biological agents yet evaluated for combatting the poisonous weed tansy ragwort *(Senecio jacobaea)* in pastures and rangelands of northwestern California, Oregon, and Washington. Three species of defoliating Lepidoptera are important supplemental agents in use against lantana in Hawaii, and a stem-mining moth, *Vogtia malloi,* is an important agent in alligatorweed control in the southeastern United States.

The Coleoptera, here construed as including the Stylopidae which are often treated as a separate order (Strepsiptera), probably contain the majority of all insects of predaceous habit. The Stylopidae are parasitic and are undoubtedly useful in regulating populations of pestiferous leafhoppers and delphacids but are not amenable to manipulation because of their complex biologies. Carabidae (ground beetles) are considered to be important in control of lepidopterous pests and one species, *Calosoma sycophanta,* is believed to be one of the important enemies of the gypsy moth in the eastern United States where it has been established.

The lady-bird beetles (Coccinellidae) are the most important of the Coleoptera as biotic agents for suppression of pests. They have been particularly useful for control of scale insects and mealybugs and to a somewhat lesser extent for aphid control. Members of one genus, *Stethorus,* feed on mites. The famous vedalia, introduced to North America to control the cottonycushion scale, is a coccinellid.

Most Coccinellidae are capable fliers and able to move about in search of suitable hosts. In fact, the tendency to disperse or migrate from areas where hosts are not available sometimes lessens their effectiveness in preventing pest outbreaks. Some species, like *Hippodamia convergens,* exhibit a consistent pattern of aggregation in the mountains in the winter and migrate from the mountains to lower elevation in the spring. These changes in flight patterns are determined by the physiological condition of the insects and the weather, and they are difficult to alter. Some success has been achieved by using food sprays (artificial honeydew) to retain beetles in the desired area. These sprays provide the necessary protein to induce oviposition when prey populations are insufficient to do so. (K.S. Hagen, E.F. Sawall, Jr. and R.L. Tassen reported on the use of food sprays to increase the effectiveness of entomophagous insects at the 1971 Tall Timbers Conference on Ecological Animal Control by Habitat Management.)

The reproductive capacity of Coccinellidae is relatively high, although the rate of development from egg to adult is not sufficiently rapid to permit beetle populations to quickly overtake a population explosion of a rapidly reproducing prey such as aphids. Fortunately, most adult beetles utilize the same prey as the larvae, although perhaps favoring a different developmental stage of the prey, and this compensates somewhat for the lagtime in population increase of the predators.

Numerous other families of beetles contain entomophagous species, but few have been specifically used as biotic agents.

Phytophagous beetles are some of the most important biological agents used for the control of weeds and were the agents of choice in the first attempt at biological control of a weed in the United States. *Chrysolina quadrigemina* and *C. hyperici,* both leaf-feeding Chrysomelidae, were introduced into the western United States in 1945–6 to control the poisonous weed *Hypericum perforatum* that then infested some 2.3 million acres of rangelands in California alone. Within a very short time the weed was effectively controlled, primarily by *C. quadrigemina,* and now remains at a level of less than 1 percent of its former abundance in California.

Another defoliating chrysomelid, *Agasicles hygrophila,* was introduced from South America in 1964 to combat the aquatic alligatorweed, *Alternanthera philoxeroides,* which then infested some 97,000 acres in eight states from Texas to North Carolina and seriously hindered navigation and other water uses. Damage to the plant by this beetle and two other introduced agents has resulted in a drastic reduction of alligatorweed abundance. Members of other families of phytophagous beetles have been used as weed control agents, particularly the weevils (family Curculionidae), which are being used to control thistles and other weedy plants.

Dung beetles (family Scarabaeidae) are being utilized in the control of flies that breed in dung, by habitat elimination, and in "control" of the dung itself.

There is a remarkable diversity of habits amongst the entomophagous Diptera (two-winged flies), and the order contains many species that are important in the biological control of pest species of insects. The great majority of the parasitic species are primary parasites of plant pests and markedly beneficial. Although some Tachinidae attack honey bees and spiders, as a whole the members of the family are beneficial. Other families which include parasitic species, such as the Bombyliidae and the Sarcophagidae, occupy an equivocal position because many are harmful while others are beneficial. Still others, such as the Nemestrinidae and the Pipunculidae, are all beneficial so far as known. Among the predaceous families, the hover flies (Syrphidae) contain numerous species that are beneficial because they prey upon aphids, scale insects and related Homoptera.

Undoubtedly, the Tachinidae is the most important family of flies for effectiveness as biocontrol agents against crop pests. Many species of many genera parasitize such important crop pests as lepidopterous larvae, larval or adult beetles belonging to several economically important groups and members of the hemipterous families Pentatomidae, Pyrrhocoridae and Coreidae. *Centeter cinerea* kills about 90 percent of the adult Japanese beetles in northern Japan when populations of that species are high and is presumed to be responsible for keeping the pest under control. *Winthemia quadripustulata* is important in suppressing armyworm populations in North America, and *Trichopoda pennipes* is similarly effective against squash bug populations and some stink bugs. *Lixophaga diatraeae, Paratheresia claripalpis* and *Metagonistylum minense* parasitize lepidopterous larvae that bore in stalks. These parasites show considerable promise as agents for control of the sugarcane borer. Several tachinids are also important parasites of the gypsy moth.

Marsh flies (Sciomyzidae) are either parasitic or predaceous on land and aquatic snails, and to a lesser degree, on slugs. Sciomyzidae that are natural enemies of snails that are intermediate hosts for helminths parasitic on man and other animals may eventually be used in the biological control of these parasitic diseases.

Some Diptera are useful in weed control. A seed fly, *Hylemya seneciella,* has been introduced into the United States to help combat tansy ragwort, and a gall fly, *Procecidochares utilis,* was imported into Hawaii to control pamakani, *Eupatorium adenophorum.* The former destroys developing seed heads; the latter, by galling the plant stems, reduces their length and foliage production to such an extent that the host weed is no longer an effective competitor. Several other anthomyiid, tephritid and cecidomyiid flies have been useful in weed control.

In general, Diptera are highly vagile and thus, they are able to seek out pest population outbreaks and, if necessary, move with mobile swarms of pests such as migratory locusts. Without their suppressive action many pest species would undoubtedly become much more serious. Unfortunately, few of the more effective species of flies are amenable to being managed in the sense that they can be readily propagated *en masse,* or to having their behavior manipulated sufficiently to assure maximal effectiveness. At present, the principal methods employed for optimizing effectiveness of flies as biotic agents are in the general category of environmental manipulation, that is, providing supplementary food sources, harborage for adults and pupation sites. These are undoubtedly the most practical means available to individual growers.

The Hymenoptera are unquestionably the dominant order among the entomophagous insects, both in the number of species having that feeding habit, and in the frequency and effectiveness with which they attack insect pests of agricultural crops and forests.

There is a fantastic diversity in the biologies of the various entomophagous species in this order, which contains most of the biotic agents routinely used in biological control actions. Among other highly specialized behavioral traits that make egg parasites among the Hymenoptera effective biotic agents is phoresy, a trait not uncommon amongst those species whose hosts lay their eggs in masses. In some groups, some species are parasitic, others predaceous and still others phytophagous, for example, species of *Eurytoma*. Some species have developed into secondary parasites (hyperparasites), that is they are parasites of a parasite. For example, a species of *Lygocerus* is parasitic on a species of *Aphidius* which is parasitic on an aphid. This habit is found only in the Hymenoptera. The development of complex biologies is common in certain groups, such as the Aphelinidae. A species of *Encarsia* parasitizes whitefly nymphs if the ovipositing female has mated, but if she is a virgin she will parasitize eggs of noctuid moths. In this case mating has the effect of altering the behavior of the female and changing the parasite-host relationship. Adult parasites in general do not depend on their host for food, but among the Hymenoptera, females of many species feed as adults on the host which they frequent, causing high mortality. For example, *Metaphycus helvolus*, a parasite of the black scale, *Saissetia oleae*, may cause higher mortality as a result of adult females feeding on the host than as a result of parasitization. Knowledge of the complex biology of parasitic Hymenoptera is essential for the successful manipulation of these species for pest control.

The preponderance of successes in classical biological control projects probably can be attributed to Hymenoptera.

Two species of parasites, *Allotropa burrelli* (Platygasteridae) and *Pseudaphycus malinus* (Encyrtidae), are credited with giving complete control of the Comstock mealybug on apple in the eastern United States. Another platygasterid species, *Amitus hesperidum,* together with *Eretmocerus serius* and some species of *Prospaltella* (Aphelinidae), has provided control of the citrus blackfly in Cuba and Mexico. *Aphytis holoxanthus* (Aphelinidae) is credited with control of the Florida red scale of citrus in Israel, and *Aphelinus mali,* another aphelinid, gives complete control of the woolly apple aphid in New Zealand. *Campsomeris annulata* (Scoliidae) gives complete control of a beetle pest of sugarcane, *Anomala sulcata,* in the Philippines. The aphelinids, *Aphytis maculicornis* and *Coccophagoides utilis,* have completely controlled the olive scale, *Parlatoria oleae,* in California. The walnut aphid, an important pest in California, has been controlled by a strain of *Trioxys pallidus* imported from Iran.

A complex of imported and native predators, together with disease organisms, is credited with substantial control of the spotted alfalfa aphid. Many hymenopterous parasites have the ability to disperse over considerable distances, thus making them particularly useful against highly mobile pest species such as the spotted alfalfa aphid. However, some parasites that lack mobility may still have the capacity to be effective agents for pest control. One such is *Neodusmetia sangwani,* which parasitizes the Rhodesgrass scale, and successfully controlled that species in south Texas after being widely distributed by airplane releases.

Spalangia endius (Spalangiidae), a parasite of the pupal stage of house flies, has been shown to have the ability to suppress housefly populations when mass produced and released in a sustained fashion. Another hymenopterous parasite, *Aphidius smithi,* has been propagated *en masse* in portable field cages in alfalfa fields for release at appropriate times to parasitize pea aphids in adjacent pea fields in the Pacific Northwest. Parasites of scale insects are also propagated in the insectary in large numbers for inoculative releases. Predaceous wasps that prey on larvae of Lepidoptera cause significant reductions of hornworm populations when colonies are concentrated in or near tobacco fields. Many other successes have been achieved by Hymenoptera of diverse families.

Autocidal Methods. Members of a species may be modified and manipulated so that they lower the reproductive or survival potential of naturally occurring populations of their own kind. Several methods have been proposed and tested: Sterilization by ionizing radiation or chemosterilants, cytoplasmic incompatibility, chromosomal translocations, introduction of deleterious or fully lethal genes and hybrid sterility. Although these differ in certain fundamentals, they have in common some basic requirements for practical application; namely, that the target pests reproduce bisexually and that there is a thorough understanding of their biology, ecological requirements, and behavior.

By far the best known of these so-called autocidal methods is the sterile-male technique. Theoretically useful for suppression of a great array of bisexually reproducing organisms, its practical application has thus far been limited to use against insect pests. In this method, large numbers of sterile insects—in most cases males only, but sometimes both sexes when sterility is complete—sufficient to greatly outnumber the resident wild population are released. A calculated proportion of ten or more sterile to one fertile insect is usually attempted. If the sterile insects are fully competitive in mating and longevity, a downward trend in the total population level will be initiated. If the release rate of sterile insects remains constant and occurs at appropriate intervals of time, the probability of fertile matings in the population will markedly decrease. Both the absolute and proportionate numbers of fertile individuals will further decline, with population extinction occurring in about four generations of the pest species, assuming effective distribution of released insects and no infiltration of fertile individuals from adjacent areas. E. F. Knipling has provided mathematical models to illustrate the population decline that may be expected.

The practicality of this method has been amply demonstrated by eradication of the screwworm fly from the southeastern United States, and its use in maintaining populations of the pest at low levels in the southwestern United States and adjacent Mexico by annual releases of 8–10 billion sterilized flies. Other uses of the technique involve the use of sterilized moths of the pink bollworm, released in the San Joaquin Valley, California, each cotton-growing season since 1967.

In a similar program, sterilized Mexican fruit flies have been released annually in southern California since 1964 to prevent the establishment of this major pest of fruits and vegetables. The release of these sterile insects has replaced the use of preventive insecticide applications. Sterilized Mediterranean fruit flies have recently been employed as a supplement to chemical sprays to eliminate an established infestation of that pest found in Los Angeles, California, in 1975. Other trials of the sterile insect method in the United States, resulting in good suppression or elimination of circumscribed natural populations, have involved populations of several pest species, namely, the boll weevil, codling moth, oriental fruit fly, melon fly, Caribbean fruit fly, cotton bollworm, horn fly, stable fly and certain species of mosquitoes.

The use of autocidal methods for biological control has certain advantages and problems. The autocidal method is absolutely specific to the target pest, and those tactics that involve releases of large numbers of altered insects, for example, sterile males or those that exhibit cytoplasmic incompatibility toward females of the wild population, have the unique property of becoming increasingly effective as the resident wild population declines, so long as the rate of releases remains constant. The ability of insects to find mates, so essential to population survival when numbers of individuals are low, becomes a disadvantage to a population subjected to sterile insect releases. Thus, the method may be especially useful for eradication of incipient or low-level infestations or to prevent low-level populations from reaching economically important levels.

Practical use of autocidal techniques requires that the following important conditions be met:

1) Suitable methods of sterilizing or genetically altering insects without chang-

ing their normal behavior or vigor, particularly as these characteristics concern mate finding and mating.

2) Availability of efficient and economical rearing methods for those tactics that involve massive releases of modified insects.

3) Reliable information regarding the population dynamics of the target pest, including quantitative information on the number and distribution of the natural population, as a guide to the number of altered insects needed in various parts of the pest's ecosystem. Initiation of autocidal methods should coincide with the predetermined low ebb in the natural pest population.

4) Efficient methods of handling and releasing the propagated insects.

5) Insects to be released that are not capable of causing annoyance or damage; if they are capable of doing so, the numbers released must be kept within tolerable limits. If chemosterilants are used on released or wild populations, there must be assurance that residues of the sterilant do not become environmental pollutants.

In summary, the principles of pest population suppression by autocidal methods are sound, and the technique provides new dimensions in pest suppression technology when used to supplement or complement other pest control methods. Under some circumstances the technique is practical when used alone, but its greatest potential is as a component in a coordinated pest management program.

Nematodes

The association between nematodes and insects varies from benign phoresy to obligate parasitism. Certain parasitic nematodes are effective biological control agents for suppression of insect pests.

Nematodes are one of the major biotic factors affecting bark beetle populations. For the most part, life histories of the parasites are synchronized with their host. Adult parasites are usually produced in adult beetles. Many of the parasites sterilize their host. It may be possible to sterilize one or both sexes of a given beetle population by the planned introduction of nematode-infested beetles, or by altering the life history of a parasite so that mature nematodes are produced in immature beetles.

The first nematode to be used for insect control, *Neoaplectana glaseri,* caused high mortality in populations of the Japanese beetle in New Jersey. More recently, another strain of this nematode has been used experimentally against a wide range of hosts, including the codling moth, the Colorado potato beetle, and the white-fringed beetle. Recently another nematode *Romanomermis culicivorax (-Reesimermis nielseni),* has been used against mosquitoes.

Augmentation of nematode populations for biological control requires mass culture procedures, preferably through use of *in vitro* techniques, which are currently unavailable. The successful rearing techniques developed for the several neoaplectanid species and mosquito parasites lend confidence that with additional research, mass culture procedures for other candidate nematodes may become available. Nematodes may be most useful against such aquatic pests as mosquitoes, black flies, and certain rice pests. Periodic inoculative releases of nematodes may be necessary for species unable to persist in host populations. In spite of a wide insect host range, there is no convincing evidence that entomogenous nematodes are a hazard to man or other vertebrates. However, many entomogenous nematodes are not host specific and may attack beneficial insects, and pathogens or toxins are associated with some nematodes; thus, careful consideration needs to be given before they are introduced into new areas.

Snails and Other Invertebrates

Two species of snails have been identified as having some potential for weed control. These are the South American snails, *Marisa cornuarietis* and *Pomacea australorbis.* The weed control potential of the *Marisa* snail was demonstrated in studies concerning its use as a biological control for the disease bilharzia. *Marisa* snails, by consuming aquatic vegetation, also consumed the eggs of the schistosome-bearing snail, *Australorbis glabratus.* Research done in Florida on the *Marisa* snail demonstrated that, under certain circumstances, the snail was capable of providing significant control of submersed aquatic weeds. However, being a tropical snail, it failed to overwinter in numbers adequate for continuous weed control even in the Miami, Florida area, and, additionally, it was found to be an indiscriminant feeder. The *P. australorbis* snail has not been studied extensively enough to provide conclusions concerning its weed control potential.

Recent studies have been made of the use of planaria worms for control of mosquitoes and chironomid midges.

Microorganisms

Animal Pathogens

The development of insect pathology as a science is relatively recent. Although known for over 2,000 years, the use of pathogens to control insects were first proposed during the 19th century. *Bacillus thuringiensis,* a potent pathogen, was discovered in 1911 and produced commercially in France in 1938. *Bacillus popilliae* was widely used for control of the Japanese beetle in the United States as early as 1939.

Relatively few pathogens have been exploited as pest control agents, and the number of potentially useful pathogens is incompletely known. For example, there are more than 500 isolates of *B. thuringiensis* that are capable of killing a broad spectrum of insect pests species; 350 nuclear polyhedrosis viruses (NPV) and granulosis viruses (GV) have been isolated from insects and mites (170). The investigation of invertebrate viruses is clearly just beginning, and the potentials of pathogens of all kinds for pest control are largely unevaluated.

There are four general kinds of pathogens considered as biological control agents, fungi, bacteria and rickettsia, protozoa and viruses. The particular biological properties of these different kinds, as well as the state of our knowledge about them, materially affect the manner in which they are used.

Fungi.   The best known of the fungi presently being used for biological control is *Beauveria bassiana.* Like most fungi, this fungus infects many different kinds of insects. It is widely used in the Soviet Union and the Peoples Republic of China where considerable success is claimed in its effectiveness against forest and orchard pests. In the Soviet Union, the combined use of the fungus with lower doses of insecticides apparently increases the efficacy of the fungus and decreases the amount of insecticide required to effect control. In France, *B. tenella* has significantly reduced numbers of soil-inhabiting larvae of the European cockchafer. Other fungi presently being studied in the United States include species active in control of the green peach aphid, the spruce budworm, and several lepidopterous pests of soybeans and cotton. In addition, fungi are being studied for use in the control of the citrus rust mite, spider mites, as well as the blueberry bud mite and the citrus bud mite. Several fungal genera, including *Coelomomyces, Lagenidium, Entomophthora* and *Culicinomyces,* show potential as control agents against the aquatic larval stages of mosquitoes and against whiteflies.

Naturally occurring fungal epidemics have materially reduced populations of some pest species, such as the seed corn maggot, by killing the adult flies. In nature, fungal epidemics depend upon rather precise conditions of temperature, humidity and host density. Thus, to use fungi effectively in biological pest control, it may be necessary to manipulate the critical physical features of the local environment or devise methods of making the fungus more tolerant.

Fungi may act as either parasites or predators in their role as biological agents for nematode control. Parasitic or endozoic fungi which attack nematodes are common in soil, but little is known about their biology. Many species do not interact with nematode pests since the spores must be ingested; therefore, they are of importance only in the population dynamics of free-living nematodes. Fungi with spores which penetrate the nematode cuticle may have some importance as regulators of pest species; but of these only one species, *Catenaria anguillulae,* has been studied, and currently its status as a biocontrol agent remains controversial. The endozoic Phycomycetes (*Haptoglossa* sp.) with nonmotile spores which penetrate the nematode cuticle, and some of the nematode-parasitic Phycomycetes appear to be important; but the obligate nature of the parasitism of most of them presents great difficulty in culturing them.

In comparison with the literature on other groups of nematode antagonists, the literature on the predaceous (nematode-trapping) fungi is extensive. Many species enmesh or ensnare nematodes by means of constricting and nonconstricting mycelial rings or by a variety of organs with adhesive surfaces or coatings. Nematodes are known to be attracted to such trapping organs. Almost all species are facultative saprophytes of varying degrees and are easily cultured. Although they can be produced in large quantities, their ability to rapidly colonize and exploit soil microhabitats is limited.

Nematode-trapping fungi are abundant and active in nature, but how abundant and under what conditions they thrive is not clear. Predaceous species of *Arthrobotrys, Dactylaria* and *Monacrosporium* are particularly abundant in both species and numbers in the rhizospheres of long-term perennial crops, and they appear to

be important factors in population dynamics of nematode pests in such crops as peach, citrus, and grapes.

Production of fungi for use as biological agents involves several considerations, such as loss of virulence when maintained on artificial media and the lack of true conidia, which are rarely produced in submerged cultures. The technology for use of fungi is not well developed or understood. Fungi are well-known allergens, and this property will need to be critically explored prior to their exploitation for use in pest control.

Bacteria and Rickettsia. The first entomopathogens to be approved for insect control and subsequent commercial production were the spore powders of *Bacillus popilliae,* active against the Japanese beetle grubs, and the formulations of the delta-endotoxin produced by *B. thuringiensis,* which are toxic to the larvae of many kinds of moths and butterflies feeding on many crops and forests. The potential of both agents has not been yet completely exploited. *Bacillus popilliae* is highly effective against the Japanese beetle, and successful development of production by submerged fermentations could greatly broaden the value of this agent. The delta-endotoxin produced by *B. thuringiensis* is widely used on lettuce, cole crops, leafy vegetables, and tobacco. It has recently been shown to be highly effective in the control of lepidopterous pests of stored grains and may replace certain chemical controls in this area. Formulations of the toxin have shown considerable promise for the control of several major forest insect pests, including tent caterpillars and webworms, gypsy moth, tussock moths, and spruce budworm. Moderate improvements in the insecticidal activities of *B. thuringiensis* formulations and application technology are needed to achieve practical and economic control of other forest pests. The toxin is also useful in controlling *Heliothis* on cotton. The HD-1 isolate of *B. thuringiensis* is used commercially in the United States, but other isolates are known to produce toxins that are more effective. In the Soviet Union, at least, four different *B. thuringiensis* formulations are used in insect control programs.

In the United States another *Bacillus* species *(B. sphaericus)* that produces a toxin that kills mosquito larvae is being studied. The bacterium may multiply and persist in the aquatic environment, can be produced in large quantities and pilot plant-sized quantities have been formulated for various field trials.

Bacteria are also useful for nematode control, although as yet not intensively exploited. Greenhouse tests using *B. penetrans* to control root lesion and root knot nematodes are very promising. In potted soil, the pathogen can destroy a population of root knot nematodes in a few generations of the host. The bacterium appears to be rather host specific. For example, an isolate from California was infective only to four species of root knot and to one species of root lesion nematodes. Nematodes not attacked by the California isolate have been reported as hosts elsewhere and it appears likely that there are biotypes of the pathogen.

Bacillus penetrans lends itself to manipulation as a biological control agent because the spore stage of its life cycle is resistant to environmental adversity. The spores fill the nematode body, causing sterility and eventually death. When the nematode disintegrates, the spores are released into the soil where they may remain viable for many years. The fact that the somewhat similar "milky spore disease" organisms of insects have been grown *in vitro* lends hope for comparable success with this nematode parasite as a microbial agent to be utilized commercially against nematode pests. No adverse effects in its use are anticipated, but an efficient means of mass-producing it must be found.

Protozoa. The use of certain protozoa, particularly microsporida, has been proposed for many years; however, few serious attempts have been made to utilize them to suppress insect pests. Several species are currently under investigation. For example, one *(Nosema locustae)* causes a disease of grasshoppers. Small plot test studies have established that distribution of 1 million spores of *N. locustae* per acre will cause a 50 percent reduction of the grasshopper population, often sufficient for effective control. About 30–50 percent of the survivors harbor the protozoan and lay fewer eggs, many of which will not be fertile. Eggs that do hatch produce diseased insects which can carry the pathogen into succeeding generations. In 1976, 16,000 acres of rangeland were treated with *N. locustae* and compared with conventional insecticide treatments and with other land left untreated as a control. No final conclusions can be drawn until the experiment is completed in 1979.

Viruses. More than 700 species of insects and several species of mites are reported to have viral diseases. Most of those affecting insects are either nuclear polyhedrosis viruses (NPV), or granulosis viruses (GV), both referred to as baculoviruses, or cytoplasmic polyhedrosis viruses (CPV). More than 320 viruses have been isolated from over 250 insect and mite species of agricultural importance, and at least 38 arthropod species have been examined in terms of control with viruses that attack them. Production of viruses from five species of insects has been industrialized on a pilot basis in the United States. Of these, the gypsy moth NPV and the Douglas-fir tussock moth NPV are of proven effectiveness as biocontrol agents. NPV's have been widely used for control of the alfalfa caterpillar in California and for control of the cabbage looper; others have been used against cotton pests in Egypt and Africa.

Plant Pathogens and Microbial Antagonists

Microbial organisms are useful both for weed control and for the control of plant pathogens.

Plant Pathogens for Weed Control. Most recent work on the use of plant pathogens to suppress or eliminate weeds centers on the use of fungi. Fungi are attractive as biological control agents because they are ubiquitous, highly host specific, destructive to the host, persistent and can be produced readily in the laboratory. Two approaches have been developed to achieve fungal control of weeds. The classical approach employs an initial inoculation of the weed with self-sustaining pathogens which require no further manipulation. The second, or mycoherbicide approach, involves an annual application of endemic or foreign pathogens in a manner similar to the use of herbicides.

The classical approach is probably better suited to control aquatic, pasture, or rangeland weeds where annual application of a pathogen for elimination of weeds is not economically feasible. Persistence and efficient natural dispersal of the pathogens, exceptionally well demonstrated by the rust fungi, are the keys to successful suppression of weeds under the classical approach. The mycoherbicide approach is better suited to annual weed control in cultivated crop lands and other areas where rapid elimination of weeds is desirable. Pathogens which can be produced *in vitro* are best suited as mycoherbicides.

The approach used may be dictated by the pathogens available, their specificity, the degree of control required, and economics. Certain pathogens will be eliminated from consideration in some situations by the above criteria.

Microbial Antagonists. Numerous biological agents exist for control of plant pathogens. Disease suppression benefits from these have been realized for many years through bioenvironmental management by use of selected cultural practices. The mechanisms of disease control in these cases are seldom understood, though research has shown that induced resistance, hyperparasitism and antagonism are the key phenomena involved. While successful examples of application of biological control on a commercial basis have been few, many others show potential.

An interaction between a pathogen and a host, which occurs in the host and makes the host more resistant to a second pathogen, is called induced resistance and is a form of biological control. Many combinations of viral, bacterial or fungal pathogens provide examples of induced resistance in the host. These may be variously called acquired resistance, cross protection, interference, and interactions. There are many mechanisms involved, but these are not well understood. Most examples have been demonstrated under laboratory and greenhouse conditions. Field results are known; for example, A.W. Helton inoculated trees with *Prunus* ringspot virus which induced resistance to *Cytosporella* cankers. Tomato seedling transplants in the Netherlands are protected from severe tobacco mosaic virus infection by resistance induced by prior inoculation with an avirulent form of the virus.

The parasitism of one parasite by another is termed hyperparasitism and appears to be a common phenomenon in nature. Its significance under field conditions is difficult to assess, and methodology for exploitation as a biological control measure has not been developed. Hyperparasitism may occur between two fungi (mycoparasitism), a bacterium and a fungus, two bacteria, a virus and bacterium, or a virus and fungus. Mycoparasites initiate parasitism either by direct penetration to the host cell or by first coiling around the host hyphae and penetrating by infection pegs. Host structures other than mycelia, such as reproductive structures, may be attacked.

A small, comma-shaped bacterium (*Bdellovibrio bacteriovorus*) is parasitic on other gram-negative bacteria. An isolate from the rhizosphere of soybean roots was used to inhibit development of another pathogen on soybeans when the two were inoculated simultaneously.

The phage-bacterium relationship may be a useful biocontrol mechanism. The phage penetrates the bacterial wall and multiplies within the protoplast, subse-quently rupturing the wall. Though it is common knowledge that phages reduce numbers of bacteria, they have not been exploited for biological control. The destructive capabilities of a virus on a fungus have been well documented by serious losses in the commercial mushroom industry because of virus diseases. Viruses have been reported in a number of pathogenic and saprophytic fungi, but their utilization for control of plant pathogens is yet to be undertaken.

Antagonists may serve as biocontrol agents by their competition for substrate or by their attempts to occupy an ecological niche and thereby interfere with pathogenicity. Organisms can also be antagonistic to each other by producing metabolic products or antibiotics. Both of these types of interaction have been used in the biocontrol of plant pathogens. Application of antagonists to seed is effective against both seed and soil-borne pathogens. Damping-off was prevented or reduced in pine, corn seedlings, and in tomato seedlings after seeds were coated with antagonistic bacteria. Some of these bacteria were shown to be antibiotic in culture to the several damping-off or seedling blight pathogens.

Protection from seedling blight was obtained by coating beet seeds with common saprophytic fungi; and oat seedlings and corn seedlings were protected by coating seeds with another common soil-inhabiting fungus. Protection against seedborne pathogens was obtained also by application of antagonists such as *Trichoderma* and *Penicillium* species to seed.

Evidence suggests that application of organisms to seed, either with or without coating materials, is feasible on a commercial scale and has the advantage of not requiring a treatment procedure by the grower. Such treatments could apply also to cuttings or other propagative plant parts. Results compare favorably with chemical seed treatments currently used in which the technology of application is similar. There is evidence also that organisms applied to seed give protection for a longer period of time than chemicals applied to seed, as organisms may multiply and maintain their inoculum potential at the seed and root surfaces. This dominance of antagonists at the seed and root surfaces may be attributed to a greater competitive saprophytic ability or to the production of antibiotic substances that deter root pathogens.

The future success of biological disease control in the phyllosphere seems to depend upon the selection of appropriate protective microorganisms and applying these to plants at a critical time under conditions which favor the antagonists. Both an increase in control effectiveness and development of economically feasible controls for field conditions are long range goals. The possibility appears remote that application of saprophytic organisms to plants might lead to undesirable features.

An interesting but poorly understood technique of biological control involves a principle that has been termed exclusive hypovirulence. Spontaneous healing of chestnut blight cankers on European chestnut was observed in several locations in Europe. Cultures of the blight fungus recovered from healing cankers were almost completely nonvirulent (hypovirulent). The ratio of these hypovirulent strains to virulent strains isolated was directly related to the health of the chestnut grove. Subsequent studies showed that active cankers could be cured by inoculation with the hypovirulent strain, and that three years after introducing the hypovirulent strain in a chestnut grove, it became dominant and the disease regressed.

If the technique of exclusive hypovirulence succeeds in the case of the chestnut blight, it may offer a significant new disease control method, providing hypovirulent strains of other pathogens can be found.

Biological control measures for diseases of forest trees and woody plants that are currently being utilized or proposed generally involve the same principle—the topical application of competing or antagonistic microorganisms to the surfaces of freshly made wounds through which certain disease- or decay-causing fungi enter. One successful biological control of a forest tree disease on a commercial scale is the use of the fungus *Phlebia (Peniophora) gigantea* to control *Fomes annosus*.

Direct biological control of a soilborne plant pathogen by a single antagonistic fungus or bacterium can be easily demonstrated in sterilized soil free of other microbial competition. However, the direct application of control agents to field soil generally has met with little success until recently. Following a report by H.D. Wells and colleagues of biological control of *Sclerotium rolfsii* by *Trichoderma harzianum* in field experiments with tomatoes, peanuts and lupines, P.A. Backman and R. Rodriguez-Kabana developed a system for growth and delivery of *T. harzianum* for field soil. The antagonist was produced in quantity on diatomaceous earth granules impregnated with a molasses solution plus potassium nitrate and potassium dehydrogen phosphate. The granules with *T. harzianum* were then mixed equally with other sterile, dry molasses-impregnated granules and distributed over-the-row to peanuts 70 and 100 days after planting. Results showed that the treatment significantly reduced pathogen incidence and damage as much as that gained by fungicide treatments; yields were also increased. The mechanism of action is thought to involve, besides antibiosis, a pH-induced competitive advantage for proteolytic enzyme production by *T. harzianum* which inhibits *S. rolfsii.*

This biocontrol system has important advantages: (1) blackstrap molasses used as a nutrient for the antagonist in the granules is inexpensive; (2) granules are lightweight and deliverable with standard machinery; (3) field results are reproducible; (4) and because *T. harzianum* is a common soil-inhabiting saprophyte, it should pose no serious environmental hazards. This method could serve as a model for delivery of other antagonists to field crops.

Biological control of certain root diseases, such as those caused by *Pythium* and *Phytophthora,* is often attributed to the presence of mycorrhizae. While plants growing in their native habitats are naturally infected with the appropriate mycorrhizal fungi, those introduced into new areas or into modified habitats such as disturbed surface areas often need to be inoculated before they are planted in the field. With relocated plants grown in nurseries, the opportunity exists to inoculate with both ectomycorrhizal or endomycorrhizal fungi, which would result not only in increased efficiency in relation to plant growth but also in protection against soil-borne pathogens. Methods for nursery inoculation of pine seedlings with mycorrhizal fungi have already been developed and might be used for other fungi.

Higher Plants as Competitors and Antagonists

Higher plants may function as biological control agents for other plants through a specialized type of interspecies competition involving the production, by some plants, of physiological chemicals that inhibit the growth of other plants. This phenomenon, called allelopathy, is well established among land plants, animals and microorganisms. It can be assumed that allelopathy is a phenomenon also active among aquatic plants. Higher plants may also be antagonistic toward plant parasitic nematodes, and some are believed to be antagonistic toward some insects.

Species of spikerush influence neighboring vegetation, which makes it possible for these diminutive plants to compete successfully with much larger species. Allelopathy is the probable basis for the successful competition. Research is in progress to further characterize the allelopathic relationship, and to extract, isolate and identify the chemical produced by spikerush. Other research is being carried on to develop practical methods by which spikerush may be used as a biological control for the larger and more troublesome species of aquatic weeds.

Allelopathic relationships are known to exist between certain agricultural crops, between weeds and crops, and between various species of weeds and other non-crop plants. Although allelopathy has been largely neglected as a useful tool in biological control of weeds, it may lead to an entirely new weed control technology.

Many plants have been tested for antagonistic properties against plant parasitic nematodes. While several plants possess antagonistic properties, crotalaria, marigold, and pangolagrass are especially effective.

Crotalaria and marigold control several of the major plant-parasitic nematodes that cause extensive losses in crop plants. Pangolagrass appears to be more restricted in scope and is effective primarily against the root knot nematode *(Meloidogyne sp.)*, but A. Ayala and others at the University of Puerto Rico reported that it controlled several other plant pathogenic nematodes as well. These plants suppress nematode populations beyond the suppression obtained by fallow, nonhost crops, or nematicides. In some cases effective control lasts more than one season.

Marigold has been used on a field scale in the Netherlands, Ceylon, Rhodesia and India to control root lesion and root knot nematodes on tea, tobacco and potato. Crotolaria has been used to control root knot nematodes in tomato transplant production by growing it after transplants are harvested and then plowing under before flowering. Pangolagrass has not yet been used commercially.

Although at least two of these plants have been used on a limited basis, none has had widespread acceptance. The reasons for limited acceptance and use are that crotolaria and marigold have no commercial value other than as green manure crops, crotalaria seeds are poisonous to livestock, marigold is difficult to establish in the field, marigold and crotalaria usually must be grown during the same season and instead of the crop to be protected, and pangolagrass must be grown for 12 to 18 months to be effective.

The future use of plants currently known to be antagonistic to nematodes will likely be limited to special circumstances. For example, marigold may be used by the home gardener who may value the ornamental aspects of the flower as well as the nematicidal benefits. Marigold may also be useful in subtropical and tropical climates where it could be grown for three to four months after the main crop has been harvested and before the next main crop is to be planted.

IPM Consultants, Firms and Associations

IPM Consulting Services

This list offers Integrated Pest Management services. Although not comprehensive, it provides an overview of the commercial activity in the IPM field.

ARIZONA

Michael Lindsey
Integrated Pest Management Services
4969 West Onyx Avenue
Glendale, AZ 85302

Mike Pursley, Manager
Safford Grower's Pest Mgmt Service
Safford, AZ 85546

CALIFORNIA

Ibrahim Michael
6747 N. Price
Fresno, CA 93727

Louis Ruud
California Agricultural Services
PO Box 525
Kerman, CA 93630

Larry Bowen
BoBiotrol
54 S. Bear Creek Drive
Merced, CA 95340

Everett Dietrick
Rincon-Vitova Insectaries
PO Box 95
Oak View, CA 93022

Maclay Burt, Executive Secretary
Association of Applied Insect
Ecologists
10202 Cowan Heights Drive
Santa Ana, CA 92705

John Nichelson
405 Oak Street
Shafter, CA 93263

Ron Inman
Agricultural Testing and Entomo-
logical
Survey
Temecula, CA 92390

ILLINOIS

David J. Harms, President
Crop Pro-Tech
33 W. Bailey Road
Naperville, IL 60540

IOWA

Laverty Sprayers, Inc.
Indianola, IA 50125

LOUISIANA

Agricultural Consulting Service,
Inc.
J.C. Patrick, President
12635 Parklake Avenue
Baton Rouge, LA 70816

Chris & Russell J. Gremillion
488 McDonald Avenue
Baton Rouge, LA 70808

Jensen Agricultural Consultants
1125 Glenmore Avenue
Baton Rouge, LA 70806

Bruce McKinnon
McKinnon's Cotton Insect Con-
trol Service
Rt. 2, Box 57
Bethany, LA 71007

Grady Coburn, President
Pest Management Enterprises,
Inc.
Box 302
Chenneyville, LA 71325

Harold J. Aymond, President
Biological-Agricultural Consult-
ants, Inc.
Box 5793
Lake Charles, LA 70606

Betty Brousseau
The Ladybugs
Box 426
Mansura, LA 71350

Elton R. Barrett
Plant Consulting, Inc.
404 Stevenson Drive
Monroe, LA 71203

J. G. Hammons
Agricultural Consulting Service
1901 Sherwood Avenue
Monroe, LA 71201

Burton R. Buckley
444 Williams Avenue
Natchitoches, LA 71457

Edward E. Jones
Jones Consultant Service
Box 726
Rayville, LA 71269

W. Henry Long
Rt. 1, Box 548-D
Thibodaux, LA 70301

Calvin Viator
Viator's Consultant Service
112 Garden Circle
Thibodaux, LA 70301

Roger R. LeBas
LeBas Pest Control & Consultant
Service
Rt 3, Box 169D
Ville Platte, LA 70586

Ray Young
Ray Young Insect Control
Box 146
Wisner, LA 71378

MISSISSIPPI

Mills L. Rogers, President
Southern Alliance of Agricultural
Consultants Association
Box 1084
Cleveland, MS 38732

D.A. Whitfield, Jr.
Whitfield Cotton Insect Service
Rt. 4, Box 146—Bayou Road
Greenville, MS 38701

NEBRASKA

Inter-American Laboratories
821 "J" Street, PO Box 94
Cozad, NE 69130

Crop Advisory Service
Elgin, NE 68636

Agricultural Consulting Service
Route 2, Box 64
Holdrege, NE 68949

Pest Management Consultants,
Inc.
3036 Prairie Road
Lincoln, NE 68506

Agricultural Technology, Inc.
McCook, NE 69001

NEW MEXICO

Bill Cox
Agricultural Technology Company
First National Tower
Las Cruces, NM 88001

Allen Lovelace
Pecos Valley Entomological Service
3013 N. Garden
Rosewell, NM 88201

SOUTH CAROLINA

Robert Moore
Agri Services Inc.
Rt 2, Box 101B
Bishopville, SC 29010

SOUTH DAKOTA

Neal Bone
Agricultural Technology Company
PO Box 526
Chamberlain, SD 57325

Tom Tzeit
Agricultural Technology Company
Pierre, SD 57501

TEXAS

Curtis Wilhelm, President
Texas Associates of Consulting
Entomologists
1609 East Jackson
Harlingen, TX 78550

WASHINGTON

Jack Eves
Agricultural Research Consultant
Rt 2, Box 221
Prosser, WA 99350

Gale Amos
Sentry Orchard Service
Rt 1, Box 304
Tieton, WA 98947

Dewey Chandler
Pest Management Consultants
1006 S. 32nd Avenue
Yakima, WA 98902

A. R. Gregorick
Sentyr Orchard Service
1516 Browne Avenue
Yakima, WA 98902

Jim Todd
Agrimanagement, Inc.
10011 South 3rd Street
Yakima, WA 98901

Beneficial Insect Firms

The following is a partial list of commercial sources that sell beneficial insects, including ladybugs, lacewings, praying mantis and trichogramma wasps, and other nonchemical controls.

CALIFORNIA

Associate Insectary
1400 Santa Paula Street
Santa Paula, CA 93040

Beneficial Insectary
2544 First Avenue
San Bernardino, CA 92405

Beneficial Insect Company
383 Waverly Street
Menlo Park, CA 94025

Bio-Control Company (Cal. Bug Co.)
Route 2, Box 2397
Auburn, CA 95603
Praying mantis, convergent lady beetles

Biosystems Fly Control
1525 63rd Street
Emeryville, CA 94608
Natural control of flies

Biotactics, Inc.
22412 Pico Street
Colton, CA 92324

Bo-Bio Control
PO Box 154
Banta, CA 95302

Bobitrol
54 South Bear Creek Drive
Merced, CA 95350

Burpee Seed Company
PO Box 748
Riverside, CA 92502
    Ladybug, praying mantis

Cal-Ag Services
480 South 4th Street
Kerman, CA 93630
    Trichogramma, green lacewings

California Green Lacewings
PO Box 2495
Merced, CA 95340

Ecological Insects Service
15075 W. California Avenue
Kerman, CA 93630

Fillmore Citrus Protective Dist.
1003 West Sespe Avenue
Fillmore, CA 93015

Fountain's Sierra Bug Company
PO Box 114
Rough & Ready, CA 95975

Harris, Paul
PO Box 1495
Marysville, CA 95901

Ladybug Sales
Route 1, PO Box 93A
Biggs, CA 95917

ORCON Organic Controls, Inc.
5132 Venice Boulevard
Los Angeles, CA 90019
    Caterpillar control, plant derivatives, *Bacillus thuringiensis*

Pyramid Nursery and Flower Shop
4640 Attawa Avenue
Sacramento, CA 95822
    Lady bugs, earthworms, lacewings

Red Leopard Ladybugs
1000 41st Street, PO Box 19008
Sacramento, CA 95819

Rincon-Vitova Insectaries, Inc.
PO Box 95
Oak View, CA 93022

Sandoz, Inc.
480 Camino del Rio South
San Diego, CA 92108
    Bacillus thuringiensis

Schnoor's Sierra Bug Company
PO Box 114
Rough & Ready, CA 95975
    Convergent lady beetles

Supervised Control Services
405 Oak Street

Shafter, CA 93263
Trichogramma, green lacewings

Unique Nursery
4640 Attawa Avenue
Sacramento, CA 95822

West Coast Ladybug Sales
Route 1, PO Box 93A
Biggs, CA 95917

CONNECTICUT

World Garden Products
2 First Street
E. Norwalk, CT 06855

FLORIDA

Pest Management Systems
Route 1, PO Box 49
Archer, FL 32618

GEORGIA

Animal Repellants, Inc.
1016 Everee Inn Road
Griffin, GA 30223
Tree tanglefoot, protective
netting

INDIANA

Galt Research, Inc.
Box 245-A, RR 1
Trafalger, IN 46181

KANSAS

Thompson-Hayward Chemical
Co.
PO Box 2383
Kansas City, KS 66110
Bacillus thuringiensis

MARYLAND

Insect Control & Research Co.,
Inc.
1330 Dillon Heights Avenue
Baltimore, Maryland 21228

MICHIGAN

The Tanglefoot Company
314 Straight Avenue SW
Grand Rapids, MI 49504
Tree tanglefoot

MINNESOTA

Ferndale Gardens
Fairibault, MN 55021

NEBRASKA

Gothard, Jack, Jr.
PO Box 7
Valentine, NE 69201

NEVADA

Pyramid Nursery
PO Box 5274
Reno, NV 89102

NEW JERSEY

Eastern Biological Control Co.
Route 5, PO Box 379
Jackson, NJ 08527

Mincemoyer Nursery
West County Line Road
Jackson, NJ 08527

NEW YORK

Butterfly Breeding Farm
389 Rock Beach Road
Rochester, NY 14617
    Praying mantis

Fairfax Biological Laboratory, Inc.
Electronic Road
Clinton Corners, NY 12514

John Staples
380 Rock Beach Road
Rochester, NY 14617
    Praying mantis

PENNSYLVANIA

Ellisco, Inc.
American & Luzerne Sts.
Philadelphia, PA 19140
    Japanese beetle bugs

King Entomological Labs
PO Box 69
Limerick, PA 19468
    Lacewings and praying mantis

Lakeland Nurseries Sales
340 Poplar Street
Hanover, PA 17331
    Ladybugs, praying mantis

Mantis Unlimited
625 Richards Road
Wayne, PA 19087

Robbins, Robert
424 N. Courtland
E. Stroudsburg, PA 18301

TEXAS

Biogenesis, Inc.
Mathis, Texas 78368
    Trichogramma

Bio Insect Control
PO Box 915
Plainview, TX 79072
    Trichogramma, green lace-
    wings, convergent lady bee-
    tle, greenbug parasites

Gothard, Inc.
PO Box 332
Canutillo, TX 79835

VIRGINIA

Reuter Laboratories
2400 James Madison Highway
Haymarket, VA 22069
    Milky spore disease

WASHINGTON

Western Biological Control Lab.
PO Box 1045
Tacoma, WA 98401
    Praying mantis, convergent
    lady beetles

CANADA
>   Fossil Flower Natural Bug Controls, Inc.
>   5266 General Road, Unit 12
>   Mississauga, Ontario L4W 127, Canada
>> Praying mantis, green lacewings, Purple Martin houses

Quarantine Facilities for Handling Biological Control Agents

There are, at present, ten quarantine facilities in operation in the United States that are authorized to handle biological control agents:

Beneficial Insects Quarantine Laboratory, Virginia Polytechnic Institute and State University, Blacksburg.

Beneficial Insect Research Laboratory, ARS, USDA, Newark, Delaware.

Biological Control Laboratory, Division of Plant Industry
Florida Department of Agriculture and Consumer Service, Gainesville.

Biological Control of Insects Research Laboratory,
ARS, USDA, Columbia, Missouri.

Biological Control of Weeds Laboratory, ARS, USDA, Albany, California.

Plant Disease Research Laboratory, ARS, USDA, Frederick, Maryland.

Quarantine Laboratory, Division of Biological Control,
University of California, Albany.

Quarantine Laboratory,
Hawaii Department of Agriculture, Honolulu.

Quarantine of Insects Research Laboratory,
ARS, USDA, Stoneville, Mississippi.

Quarantine Laboratory, Division of Biological Control,
University of California, Riverside.

Descriptions of Private IPM Firms

As part of the information-gathering effort for this book, we surveyed a number of firms offering IPM services to farmers. The following questions were asked:

1. What are the major crops grown by participating farmers?
2. What crop pests are included in your IPM program?

3. What are the major pest control methods used in your IPM program?

4. Who determines your IPM program, how many people are employed in your firm to administer IPM services, and what kinds of educational backgrounds do these employees have?

The following selection illustrates the range of replies we received.

Rincon-Vitova Insectaries, Inc.
Oakview, California
Everett J. Dietrick, President

Integrated pest management is useful in all crops to some extent. We try to talk to the farmer customers about all their crops, since in this way we can help him to farm with least interference to the natural enemy complexes that are so important in biological control. A positive effort is made to augment the populations of good bugs sufficiently to prevent the need for pesticides wherever possible.

1. Major crops: Cotton, citrus, avocadoes, pears, apples, corn, walnuts, almonds, stone fruits, tomatoes, strawberries, olives, grapes, (wine & table), eggplant, artichokes, lettuce, potatoes, geraniums, greenhouses, fly control on poultry and egg farms, feedlots and dairies, alfalfa hay and seed, sorghum and other grains, bell and chili peppers, ornamentals and tree farms, sugar beets, broccoli and cabbage carrots and others

2. Major crop pests: Cotton bollworms, pink bollworm, boll weevil, lygus, thrips, fleahoppers, cotton leaf perforator, California red scale, citrus thrips, citrus red spider, black scale and brown soft scale, Latania scale, avocadoes loopers etc. Avocado brown mite, codling moth, pear psillid, corn earworm corn borer, walnut husk fly, navel orange worm, walnut aphids, two-spotted mites grape leafhopper, potato tuber moth, cabbage looper, peach twig borer, oriental fruit moth, green house whitefly, green peach aphid, artichoke plume moths, olive scales, beet army worms, yellow stripped army worms, the army worms, alfalfa butterfly, alfalfa weevils, spotted alfalfa and blue-green aphids, green bug, grain aphids, pepper weevil, manure inhabiting flies ie. Musca spp and Fannia spp., aphids on ornamentals ie.Spirea aphid etc., cabbage aphid, strawberry aphids and mites.

3. Major pest control methods: The principal objective is to develop the natural enemy complex that manages the pest complexes associated with each crop or group of crops. Habitat management and cultural controls are exploited to the fullest, crop diversity is expanded using cover crops, insectary crops, trap crops etc. Mulching and other biological farming techniques are suggested along with green manure cropping. Food chains are developed based on decomposer groups using sheet composting to form habitats that keep predators active on the farms. Selective pesticides and some other pesticides used selectively form the basis of guarantee when all else fails and no predictable biological controls are possible.

4. The individual PCA (pest control advisor) who watches the fields calls the shots on his growers' farms. Since IPM is more a set of principles and philosophy than any program, we depend on the advisor making the decisions for each field based on the best sampling information and ecology of the situation.

All our employees are college graduates, specializing in at least biological sciences, mostly entomology. We presently have ten full time field people.

Jensen Agricultural Consultants, Inc.
Baton Rouge, Louisiana
Richard L. Jensen, Ph.D.

Since there are a diversity of crops grown in Louisiana, many consultants tend to specialize in one major crop. Thus, we have consultants who deal mainly with cotton, others with sugarcane and others with rice and soybeans. However, there are areas in Louisiana where many crops are grown and those of us who work in these areas consult on several crops. For example, last season I consulted on sugarcane, rice, cotton, soybeans, sorghum, bermudagrass and corn. There are also many vegetables grown in my area, but my organization is just not large enough at the present time to handle crops such as sweetpotatoes, peppers, cabbage and tomatoes. I also do a considerable amount of consulting with larger nurseries. The remainder of my time is consumed doing contract research for various chemical companies.

I have decided to show growers what can be done by using an effective insect pest management system. I also integrate portions of various pest management systems developed for different crops. For example, the southern green stink bug is considered the major insect pest of soybeans. The first generation of stink bugs develops on clover, the second on corn and rice, and the third, and possibly fourth, on soybeans. One of the components of the present soybean pest management system recommended by Louisiana State University involves the use of a trap crop. Since stink bugs are attracted to early maturing

soybean varieties, we recommend that a grower plant small, strategically located plots of soybeans 10–14 days earlier than the adjacent main crop. Then as the stink bugs invade the trap plots, they are controlled with chemical insecticides. Although this concept requires treatment of stink bug populations that are below the economic injury threshold, the acreage treated (one to five percent) is such a small portion of the entire crop that predators from the adjacent areas readily reinvade the trap plots and outbreaks of secondary pests are held to a minimum.

With farmers that produce both corn and soybeans, I incorporate management systems from two crops and use corn as a trap crop for soybeans. Because large numbers of stink bugs will congregate on the borders of corn fields, these populations can be controlled by treating the borders of the field. Thus, I am able to decrease the second generation rather than the third generation of stink bugs. This delays the buildup of southern green stink bugs in soybeans and often the population does not reach the economic injury threshold.

As an agricultural consultant, I believe that we should go beyond IPM and deal more with Farm Management. There are many factors other than pests that can reduce yield. Herbicides applied using improper nozzles, fertility of the soil and proper drainage of the land are just a few.

Pest Management Consultants, Inc.
Lincoln, Nebraska

1. Major Crops: Dent corn, popcorn and seed corn. Milo (grain sorghum), soybeans and alfalfa.
2. Crop pests include: Corn rootworms, cutworms, European corn borer, Western

bean cutworm, corn earworm, spider mites (banks grass mite and two spotted mite), grasshoppers, armyworm, greenbug, corn leaf aphid, alfalfa weevil, potato leafhopper, lygus bugs, thrips.

3. Pest control methods: crop rotation, resistant and tolerant varietal selection, planting times, cultural practices, monitoring parasite, predator and disease populations and incidence, chemicals.

4. IPM programs are determined by three professionals. Dr. Earle S. Raun heads the firm and has a Ph.D. in Entomology and 29 years of experience. Dr. Lester Leininger has a Ph.D. in Agronomy (major crop genetics, minor soils) and 22 years experience. Bruce Monke has an M.S. in Entomology (completing his Ph.D.) and has three years experience.

PMC also offers services in soil fertility and water management.

Agriservices, Inc.
Bishopville, South Carolina

1. Major crops: soybeans, cotton, corn and tobacco.

2. Crop pests: Cotton—tobacco budworm, cotton bollworm, boll weevil, armyworms, lygus, aphids, thrips, mites. Soybeans—Heliothis Spp., stinkbugs, green clover worm, velvetbean cat., Spodoptera spp., soybean looper, Mexican bean beetle.

3. Major pest control methods: IPM scouting program: *Proper identification.* Using information such as life histories, behavior, ecology and other important factors. *Adequate surveying.* Determining pests present and whether control measures are actually needed. *Utilization of Economic Thresholds and Economic Injury Levels.* Agriservices, Inc. is not solely interested in in-sect suppression, but insect suppression which is profitable. Old ideas for maximum yields employed strict schedules; however, we work on an as-needed basis spray program.

4. Full-time are Robert E. Moore (Masters, Entomology) and one employee with B.S., Biology. Part-time summer employees are college students.

Pecos Valley Entomological Service
Roswell, New Mexico

1. Major crops: cotton, alfalfa, corn (ensilage), sorghum (milo), sunflowers, pecans.

2. Crop pests: Lygus, bollworm, cabbage looper, pink bollworm.

3. Pest control methods (in order): Cultural controls, evaluation of beneficial populations, early harvest, spray with microbial insecticides, spray with standard insecticides.

4. The farmer is paying for our advice and may do as he pleases concerning our IPM program except for planting/plow-up dates when cultural controls are in effect for pink bollworm. Four adults in the two field crews have a Master's degree.

Agricultural Consulting Service
Holdrege, Nebraska

1. Major crops: corn, soybeans, alfalfa, sorghum.

2. Primary crop pests: corn rootworm, corn borer, spider mite, cutworm, Western bean cutworm, greenbug, aphids and weevils.

3. Major pest control methods: chemical spraying—ground or aerial; rotation and variety selection, such as for greenbug in sorghum and, to a limited extent, for corn borer.

4. As owner of company, Eugene J. Schwartz (agronomist) determines IPM program. Four full-time staff with M.S. and B.S. degrees. In 1977, firm hired several field scouts who had at least two years of college, with two working on M.S. degrees.

Agricultural Technology Company
McCook, Nebraska
and
Tempe, Arizona

1. Major crops: corn, sugar beets, pinto beans, soybeans, sorghum alfalfa, barley and wheat.
2. Pests: insects, weeds and diseases.
3. Methods: Crop rotation, chemicals and varieties.
4. IPM program is determined by both staff and help from state universities. Firm employs approximately 40 people to administer services. Backgrounds generally include B.S. and M.S. degrees in agronomy, entomology, crop production or crop protection.

Integrated Pest Management Systems
Glendale, Arizona

1. Major crops: cotton, small grains, sugar beets, safflower, alfalfa.
2. Crop pests: pink bollworm, lygus, cotton bollworm, cabbage looper, beet armyworm, greenbug, Egyptian alfalfa weevil, cotton leaf-perforator.
3. Methods: natural, cultural and chemical controls.
4. Michael L. Lindsey determines the kind

of IPM program that is put into effect (usually a combination of above controls). The number of employees is determined by acreage to be checked. The educational background of employees varies from juniors in high school to school teachers, college students or graduates.

Crop Pro-Tech
Naperville, Illinois

1. Major crops: corn and beans.
2. Major pests: Anything that may hinder production.
3. Methods: Mechanical control, cultural practices, biological and chemical controls are all considered to be viable.
4. David J. Harms is president. There are six employees, 4 Ph.D.'s, and two others. They have 12 to 15 years experience each. Backgrounds include economics, soil science, plant pathology, weed science and general agronomy experience.

Laverty Sprayers, Inc.
Indianola, Iowa

Now in third year of a pest management program for corn rootworm beetles. Program is offered to corn growers through agricultural chemical dealers. An insect scout monitors and evaluates corn rootworm beetle population to determine if and when a control program is needed. Listed costs are $2.00 per acre—beetle scouting only (no chemical treatment); $7.75 per acre—beetle scouting with one chemical treatment; $13.50 per acre—beetle scouting with two chemical treatments.

Government Cooperation With Private Companies

Following are some examples of cooperation between the United States Department of Agriculture and State Cooperators with the Private Sector in the development and implementation of pest management tactics and systems. The information is provided by R. L. Ridgway, National Program Staff, Federal Research of the USDA, Beltsville Agricultural Research Center in Beltsville, Maryland.

Tactics or systems	Locations providing primary contributions	Principals involved in implementation
(1) Improved strains of *Bacillus thuringiensis* (B.t.)	Brownsville, TX	Abbott & Sandoz, Inc.
(2) Predictive modeling of *Heliothis*	College Station, TX	Light trap manufacturers, TX Agr. Ext. Serv.
(3) *Heliothis* NPV	Fayetteville, AR Brownsville, TX College Station, TX	Sandoz, Inc.
(4) Gossyplure and mating disruption of pink bollworm	Riverside, CA College Station, TX National Program Staff	Conrel and Wilbur-Ellis
(5) Insect growth regulator, Dimilin	Florence, SC College Station, TX Raleigh, NC	Thompson-Hayward Chemical Co.
(6) *Trichogramma* egg parasite	College Station, TX	Biogenesis, Inc. Gothard, Inc. Rincon-Vitova, Inc.
(7) IPM system (short-season cotton production)	Weslaco, TX Brownsville, TX	Seed producers TX. Agr. Ext. Serv.
(8) IPM system (B.t. and/or *Trichogramma,* and improved decision-making technology)	Stoneville, MS College Station, TX	Abbott, Cotton Inc., private consultants, MS. Agr. Ext. Serv.
(9) IPM system (gossy-plure, *Trichogramma,* and improved decision-making technology)	Phoenix, AZ College Station, TX	Wilbur-Ellis

Suggested Reading List

————. *Plant Disease Handbook.* 3rd ed. New York: Van Nostrand Reinhold, 1971.

Askew, R.R. *Parasitic Insects.* New York: Elsevier, 1971.

Baker, K. F. and W. C. Snyder, eds. *Ecology of Soil-Borne Plant Pathogens.* Berkeley: University of California Press, 1965.

Baker, Kenneth F. and R. James Cook. *Biological Control of Plant Pathogens.* San Francisco: W. H. Freeman and Co., 1974.

Barbosa, P. and T.M. Peters. *Readings in Entomology.* Philadelphia: Saunders, 1972.

Beal, G.M., J.M.Bohlen, and H.G. Lingren. *Behavior Studies Related to Pesticides Agricultural Chemicals and Iowa Farmers.* Special Report No. 49, Ames, Iowa: Iowa State University of Science & Technology.

Beirne, B.P. *Pest Management.* London: Leonard Hill Books, 1969.

Bird, G.W. *On-Line Pest Management.* E. Lansing: Michigan State University, 1977.

Bottrell, D.G., C.B. Huffaker and R.F. Smith. *Information Systems for Alternate Methods of Pest Control.* UC/AID Pest Management and Related Environmental Protection Project Report. Berkeley: University of California, 1976.

Brooklyn Botanical Garden. *Handbook on Biological Control of Plant Pests.* New York: BBG.

Brown, A.W., T.C. Byerly, M. Gibbs and A. San Pietro, eds. *Crop Productivity—Research Imperatives.* Michigan State University, Agricultural Experiment Station and Charles F. Kettering Foundation, 1975.

Bruehl, G.W., ed. *Biology and Control of Soil-Borne Plant Pathogens.* St. Paul: American Phytopathological Society, 1975.

Burges, H.D. and N.W. Hussey, eds. *Microbial Control of Insects and Mites.* New York: Academic Press, 1971.

California Agriculture. Special Issue: Integrated Pest Management. Vol. 32, No. 2. Division of Agricultural Sciences. University of California, Berkeley, February, 1978.

Carson, R. *Silent Spring.* Greenwich, CT: Fawcett, 1962.

Carter, W. *Insects in Relation to Plant Disease.* New York: Wiley, 1973.

Casida, J.E., ed. *Pyrethrum, The Natural Insecticide.* New York: Academic Press, 1973.

Christie, J.R. *Plant Nematodes: Their Bionomics and Control.* Gainesville: University of Florida, 1959.

Clark, L.R., P.W. Geier, R.D. Hughes and R.F. Morris. *The Ecology of Insect Populations in Theory and Practices,* 1967.

Clausen, C.P. *Entomophagous Insects.* New York: McGraw-Hill Book Co., Inc., 1940.

Clausen, C.P. *Biological Control of Insect Pests in the Continental United States.* U.S. Department of Agriculture, Technical Bulletin No. 1139, 1956.

Clausen, C.P., ed. *Introduced Parasites and Predators of Arthropod Pests and Weeds: A World Report.* U.S. Department of Agriculture, Agricultural Handbook 480, 1977.

Coppel, H.C. and J.W. Mertins. *Biological Insect Pest Suppression.* Verlay, New York: Springer Co.

Corbett, P.S. and J.S. Kelleher, P. Harris, W. Reeks, G.D. Williamson and R.M. Prentice (compilers). *Biological Control Programmes Against Insects and Weeds in Canada, 1959–1968.* Commonwealth Institute of Biological Control Technical Communication, 1971.

Council on Environmental Quality. *Integrated Pest Management.* Washington, D.C.: U.S. Government Printing Office, 1978 (Revised).

Day, P. *Genetics of Host-Parasite Interaction.* San Francisco: W.H. Freeman, Inc., 1974.

DeBach, P., ed. *Biological Control of Insect Pests and Weeds.* New York: Reingold Publication Corp., 1964.

DeBach, P. *Biological Control of Natural Enemies.* New York: Cambridge University Press, 1974.

Delucchi, V.L., ed. *Studies in Biological Control.* Cambridge: Cambridge University Press, 1976.

Elton, C.S. *The Ecology of Invasion by Animals and Plants.* London: Methuen and Co., Ltd., 1958.

Fleming, W.E. *Biology of the Japanese Beetle.* U.S. Department of Agriculture, Agricultural Research Service, Technical Bulletin 1449, 1972.

Flint, M.L. and R. van den Bosch. *A Source Book on Integrated Pest Management.* Department of Health, Education and Welfare, Office of Education, Office of Environmental Education, 1977.

Food and Agriculture Organization (FAO). *Proceedings of the FAO Symposium on Integrated Pest Control.* 3 vols. Rome: FAO, 1966.

Gerberich, J.B. and M. Laird. *Bibliography of Papers Related to the Control of Mosquitoes by the Use of Fish (an annotated bibliography for the years 1901–1966).* FAO Fisheries Technical Paper No. 75. Rome: FAO United Nations, 1968.

Glaser, R.W. *Studies on "Neoaplectana glaseri," a nematode parasite of the Japanese beetle (Popillia Ojaponica).* New Jersey Department of Agriculture Circular No. 211, 1932.

Good, J.M. et al. *Establishing and Operating Grower-Owned Organizations for Integrated Pest Management.* Extension Service. Washington, D.C.: U.S. Department of Agriculture, 1977.

Guenzi, Wayne D., ed. *Pesticides in Soil and Water.* Madison, Wisconsin: Soil Science Society of America.

Hairston, N.G., K.E. Wurzinger and J.B. Burch. *Non-Chemical Methods of Soil Control.* World Health Organization, WHO/VBC/75.573, 1975.

Hepp, R.E. *Alternative Delivery Systems for Farmers to Obtain Integrated Pest Management Services.* Michigan State University. Agricultural Economics Report 298 1976.

Hills, L.D. *Pest Control Without Poisons.* Essex, England: Henry Doubleday Res. Assoc., Bocking, Braintree, 1964.

Huffaker, C.B., ed. *Biological Control.* New York: Plenum Press, 1971.

Huffaker, C.B. and P.S. Messenger, eds. *Theory and Practice of Biological Control.* New York: Academic Press, 1976.

Hunter, B. *Gardening Without Poisons.* 2d ed. Boston: Houghton-Mifflin, 1971.

International Union Forest Research Organizations. *Biological Control of Forest Diseases.* XV Congress. Gainesville, 1971.

Jacobson M. and D.G. Crosby, eds. *Naturally Occurring Insecticides.* New York: Decker, 1971.

Johnson, P.T. *An Annotated Bibliography of Pathology in Invertebrates Other Than Insects.* Minneapolis: Burgess Publishing Co., 1968.

Johnson, P.T. and F.A. Chapman. *An Annotated Bibliography of Pathology in Invertebrates Other Than Insects.* Supplement. Miscellaneous Publication No. 1. University of California, Irvine: Center for Pathobiology, 1969.

Jones, D.P. and M.E. Solomon, eds. *Biology in Pest and Disease Control.* 13th Symposium of the British Ecological Society. Oxford: Blackwell Scientific Publication, 1974.

Jones, S.L. and R.L. Ridgway. *Development of Methods for Field Distribution of Eggs of the Insect Predator "Chrysopa carnea" Stephens.* U.S. Department of Agriculture, Agricultural Research Service, ARS-S-124, 1976.

Kilgore, W.W. and R.L. Doutt, eds. *Pest Control: Biological, Physical and Selected Chemical Methods.* New York: Academic Press, 1967.

Knipling, E.F. *The Potential Role of the Sterility Method for Insect Population Control With Special Reference to Combining this Method With Conventional Methods.* U.S. Department of Agriculture, Agricultural Research Service, ARS 33–98, 1964.

Knorr, L.C. *Citrus Diseases and Disorders.* Gainesville: University Press of Florida, 1973.

Komarek, E.V. *Proceedings of the Tall Timbers Conference on Ecological Animal Control by Habitat Management.* Tallahassee, Florida: Tall Timbers Research Station, 1969.

Kovalev, D.V. *(The Use of a Biological Method of Controlling Weedy Plants).* (In Russian). Obzornaia Informatsiia (Moscow) 749., 1973.

Kuenen, D.J., ed. *The Ecological Effects of Biological and Chemical Control of Undesirable Plants and Animals.* 8th Technical Meeting of the International Union for Conservation of Nature and Natural Resources, Warsaw. Leiden, Netherlands: E.J. Brill, 1961.

Mankau, R. *Utilization of Parasites and Predators in Nematode Pest Management Ecology. Proceedings.* Tall Timbers Conference on Ecological Animal Control by Habitat Management 4, 1973.

Martignoni, M.E. and P.J. Iwai. *A Catalog of Viral Diseases of Insects and Mites.* U.S. Department of Agriculture, Forest Service Technical Report PNW-40, 1975.

Massey, C.L. *Biology and Taxonomy of Nematode Parasites and Associates of Bark Beetles in the United States.* U.S. Department of Agriculture, Agricultural Handbook No. 446, 1974.

Maxwell, F.G. and F.A. Harris, eds. *Proceedings of the Summer Institute Biological Control of Plant Insects and Diseases.* Jackson: University Press of Mississippi, 1974.

McCabe, Robert, et. al., eds. *Integrated Pest Management: A Curriculum Report.* U.S. Office of Education. Office of Environmental Education, 1973.

McLeod, J.H., B.M. McGugan and H.C. Coppel. *A Review of the Biological Control Attempts Against Insects and Weeds in Canada.* Commonwealth Institute of Biological Control Technical Communication 2., 1962.

Metcalf, C.L., W.P. Flint, and R.L. Metcalf. 4th ed. *Destructive and Useful Insects, Their Habits and Control.* New York: McGraw-Hill Book Co., 1962.

Metcalf, R.L. and W.H. Luckman, eds. *Introduction to Insect Pest Management.* New York: Wiley & Sons, 1975.

Michigan State University. *Ecosystem Design and Management II Integrated Pest Management: An On-Line Control Process.* Washington, D.C.: National Science Foundation, 1974.

Miller, P.M. and J.F. Ahrens. *Marigolds—A Biological Control of Meadow Nematodes in Gardens.* Connecticut Agricultural Experiment Station Bulletin 701, 1969.

Mock, D.E. *Pest Management and the Development of Field Crop Consulting Services in Kansas.* Manhattan: Kansas State University, 1976.

Mrak, E.M. (Chairman). *Report of the Secretary's Commission on Pesticides and Their Relationship to Environmental Health.* U.S. Department of Health, Education and Welfare. Washington, D.C.: U.S. Government Printing Office, 1969.

National Academy of Sciences. eg. *Principles of Plant and Animal Pest Control.* Washington, D.C.: NAS, 1968–1969. 1968 Vol. 1—*Plant-Disease Development and Control.* 1968 Vol. 2—*Weed Control.* 1969 Vol. 3—*Insect-Pest Management and Control.*

National Academy of Sciences. *Pest Control Strategies for the Future.* Washington, D.C., 1972.

National Academy of Sciences. *Pest Control: An Assessment of Present and Alternative Technologies.* Vol. I *Contemporary Pest Control Practices and Prospects: The Report of the Executive Committee.* NAS, Washington, D.C., 1975. Vol. II—*Corn/Soybeans Pest Control.* Vol. III—*Cotton Pest Control.* Vol. IV—*Forest Pest Control.* Vol. V—*Pest Control and Public Health.*

National Research Council. *Insect-Pest Management and Control.* Washington, D.C. No. 1695, 1969.

Oborn, E.I., W.T. Moran, K.T. Greene, and T.R. Bartley. *Weed Control Investigations on Some Important Aquatic Plants Which Impede Flow of Western Irrigation Waters.* U.S. Department of Interior, Bureau of Reclamation—U.S. Department of Agriculture, Agricultural Research Service, Joint Laboratory Report S1-2. Denver: Bureau of Reclamation, 1954.

Olkowski, Helga. *Common Sense Pest Control.* Richmond, California: Consumers Cooperative of Berkeley, Inc., 1971.

Ordish, G. *Biological Methods in Crop Pest Control: An Introductory Study.* London: Constable, 1967.

Painter, Reginald H. *Insect Resistance in Crop Plants.* Lawrence: University Press of Kansas, 1951.

Papavizas, G.C., ed. *The Relation of Soil Microorganisms to Soil-borne Plant Pathogens.* Southern Cooperative Series Bulletin 183. Blacksburg: Virginia Polytechnic Institute and State University, 1974.

Pennsylvania State University. *Disease and Insect Problems in the Commercial Orchard.* University Park: Penn State University.

Philbrick, J. and H. Philbrick. *The Bug Book: Harmless Insect Controls.* Charlotte, VT: Garden Way, 1963.

Pimentel, David. *Ecological Effects of Pesticides on Nontarget Species.* Executive Office of the President, Office of Science and Technology. Washington, D.C.: U.S. Government Printing Office, 1971.

Pimentel, David., ed. *Pest Control Strategies— Understanding and Action.* New York: Academic Press, 1978.

Poinar, G.O., Jr. *Entomogenous Nematodes.* Leiden, Netherlands: E.J. Brill, 1975.

Preece, T.F. and C.H. Dickinson., eds. *Ecology of Leaf Surface Microorganisms.* New York: Academic Press, 1971.

Quraishi, Sayeed M. *Biochemical Insect Control: Its Impact on Economy, Environment and Natural Selection.* New York: John Wiley & Sons, 1977.

Rabb, R.L. and F.E. Guthrie, eds. *Concepts of Pest Management.* Raleigh: North Carolina State University Press, 1970.

Rees, N.E. *Arthropod and Nematode Parasites, Parasitoids and Predators of Acrididae in America North of Mexico.* U.S. Department of Agriculture, Technical Bulletin 1460, 1973.

Regev, V., A. Gutierrez and G. Feder. *Pests as a Common Property Resource: A Case Study in the Control of the Alfalfa Weevil.* Draft Manuscript. Department of Agricultural and Resource Economics, University of California, Berkeley, 1975.

Ridgway, R.L. and S.B. Vinson, eds. *Biological Control by Augmentation of Natural Enemies: Insect and Mite Control with Parasites and Predators.* New York: Plenum Press, 1977.

Rudd, R.L. *Pesticides and the Living Environment.* Madison: University of Wisconsin Press, 1964.

Rudd, Robert L. *Pesticides and the Living Landscape.* Madison: University of Wisconsin Press, 1966.

Sabrosky, C.W. and R.C. Reardon. *Tachinid Parasites of the Gypsy Moth, "Lymantria Dispar," With Keys to Adults and Puparia.* Entomological Society of America Miscellaneous Publications 10 (2), 1976.

Smith, C.N., ed. *Insect Colonization and Mass Production.* New York: Academic Press, 1965.

Smith, L.B. and C.H. Hadley. *The Japanese Beetle.* U.S. Department of Agriculture, Agricultural Research Service, Circular 363, 1926.

Stevens, R.E. and F.G. Hawksworth. *Insects and Mites Associated with Dwarf Mistletoes.* U.S. Department of Agriculture, Forest Service, Rocky Mountain Forest and Range Experiment Station Research Paper RM-59. Fort Collins, 1970.

Stone, J.C. *Pesticides and Your Environment: A Guide for the Homeowner and Home Gardener.* Washington, D.C.: National Wildlife Federation, 1972.

Suatmadjii, R.W. *Studies on the Effect of "Tagetes" Species on Plant Parasitic Nematodes.* Wagoningen: H. Veenman and Zonen, N.V., 1969.

Summers, M., R. Engler, L.A. Falcon, and P. Vail., eds. *Baculoviruses for Insect Pest Control: Safety Considerations.* Washington, D.C.: American Society of Microbiology, 1975.

Swan, L. *Beneficial Insects.* New York: Harper and Row, 1964.

Tamaki, G. and R.E. Weeks. *Biology and Ecology of Two Predators, "Geocoris Pallens" Stal and "G Bullatus" (Say).* U.S. Department of Agriculture, Agricultural Research Service, Technical Bulletin 1446, 1972.

Targan, A.C. and B.E. Hopper. *Nomenclatorial Compilation of Plant and Soil Nematodes.* Society of Nematology, Inc. De Leon Springs, Florida: E.O. Painter Printing Co., 1974.

Townes, H. *The Genera of Ichneumonidae, Part I.* Ann Arbor, MI: The American Entomological Institute, 1969.

Tummala, R.L. et al. *Modeling for Pest Management.* E. Lansing: Michigan State University, 1974.

U.S. Army Corps of Engineers. *Herbivorous Fish for Aquatic Weed Control.* U.S. Army Engineers Waterways Experiment Station, Vicksburg, Aquatic Plant Control Program, Technical Report 4, 1973.

U.S. Environmental Protection Agency. *Impact of the Use of Microorganisms on the Aquatic Environment.* Ecological Research Series, EPA 660/3-75-001. Corvallis, Oregon.

van den Bosch, Robert. *The Pesticide Conspiracy.* New York: Doubleday Co., 1978.

van den Bosch, Robert and P.S. Messenger. *Biological Control.* New York: Intext Educational Publishers, 1973.

van den Bosch, Robert et. al. *Investigation of the Effects of Food Standards on Pesticide Use.* Environmental Protection Agency, Contract No. 68-01-2602.

von Rumker, Rosmarie. *Farmers' Pesticides Use Decisions and Attitudes on Alternate Crop Protection Methods.* Office of Pesticide Programs. Office of Water and Hazardous Materials. Environmental Protection Agency, 1974.

von Rumker, R., R.M. Matter, D.P. Clement and F.K. Erickson. *The Use of Pesticides in Suburban Homes and Gardens and Their Impact on the Aquatic Environment.* Pesticide Study Series 2. Environmental Protection Agency Office of Water Programs Applied Technology Division. Washington, D.C.: U.S. Government Printing Office, 1972.

von Rumker, R. and Gary L. Kelso. *A Study of the Efficiency of the Use of Pesticides in Agriculture.* Office of Pesticide Programs, Environmental Protection Agency, 1975.

von Rumker, R., G.A. Carlson, R.D. Lacewell, R.B. Norgaard, and D.W. Parvin, with F. Horay, J.E. Casey, J. Cooper, A.H. Grube and V. Ulrich. *Evaluation of Pest Management Programs for Cotton, Peanuts and Tobacco in the United States.* U.S. Environmental Protection Agency, EPA 540/9-75-031, 1975.

Wardle, R.A. and P. Buckle. *The Principles of Insect Control.* Manchester, England: Manchester University Press, 1923.

Webster, J.M. ed. *Economic Nematology.* New York: Acadmeic Press, 1972.

Westphal, Karl. *The 1977 Seminar on Cockroach Control.* N.Y. State Department of Health, 1977.

Willey, W.R.Z. *The Diffusion of Pest Management Information Technology.* Ph.D. Dissertation, Department of Economics. University of California, Berkeley, 1974.

Wilson, H.R., B.L. Damron and R.H. Harms. *Geese Vs. Water Hyacinths.* Sunshine State Agricultural Research Report, Spring, 1977.

Wood, D.L. R.M. Silverstein and M. Nakajima, eds. *Control of Insect Behavior by Natural Products.* New York: Academic Press, 1970.

Woods, A. *Pest Control.* New York: John Wiley & Sons, 1974.

World Health Organization. *Insecticide Resistance and Vector Control.* Technical Report Series 443, 1970.

World Health Organization. *Conference on the Safety Biological Agents for Arthropod Control.* WHO Report VBC/73.1, 1973.

World Health Organization (WHO). *The Use of Viruses for Control of Insect Pests and Disease Vectors.* WHO Technical Report Series 531, 1973.

Zuckerman, B.M., W.F. Mai and R.A. Rohde, eds. *Plant Parasitic Nematodes.* Vol. 1. New York: Academic Press, 1971.

Index